I0756440

The Michigan Historical Reprint Series

Reprints from the collection of the University of Michigan University Library

This volume is produced from digital images created through the University of Michigan University Library's preservation reformatting program. The Library seeks to preserve the intellectual content of items in a manner that facilitates and promotes a variety of uses. The digital reformatting process results in an electronic version of the text that can both be accessed online and used to create new print copies. This book and thousands of others can be found in the digital collections of the University of Michigan Library. The University Library also understands and values the utility of print, and makes reprints available through its Scholarly Publishing Office.

For access to the University of Michigan Library's digital collections, please see http://www.lib.umich.edu.

The Scholarly Publishing Office seeks to disseminate high-quality, cost-effective scholarly content through both print and electronic publishing. Information about the Scholarly Publishing Office can be found at http://spo.umdl.umich.edu.

The Scholarly Publishing Office
the University of Michigan
University Library

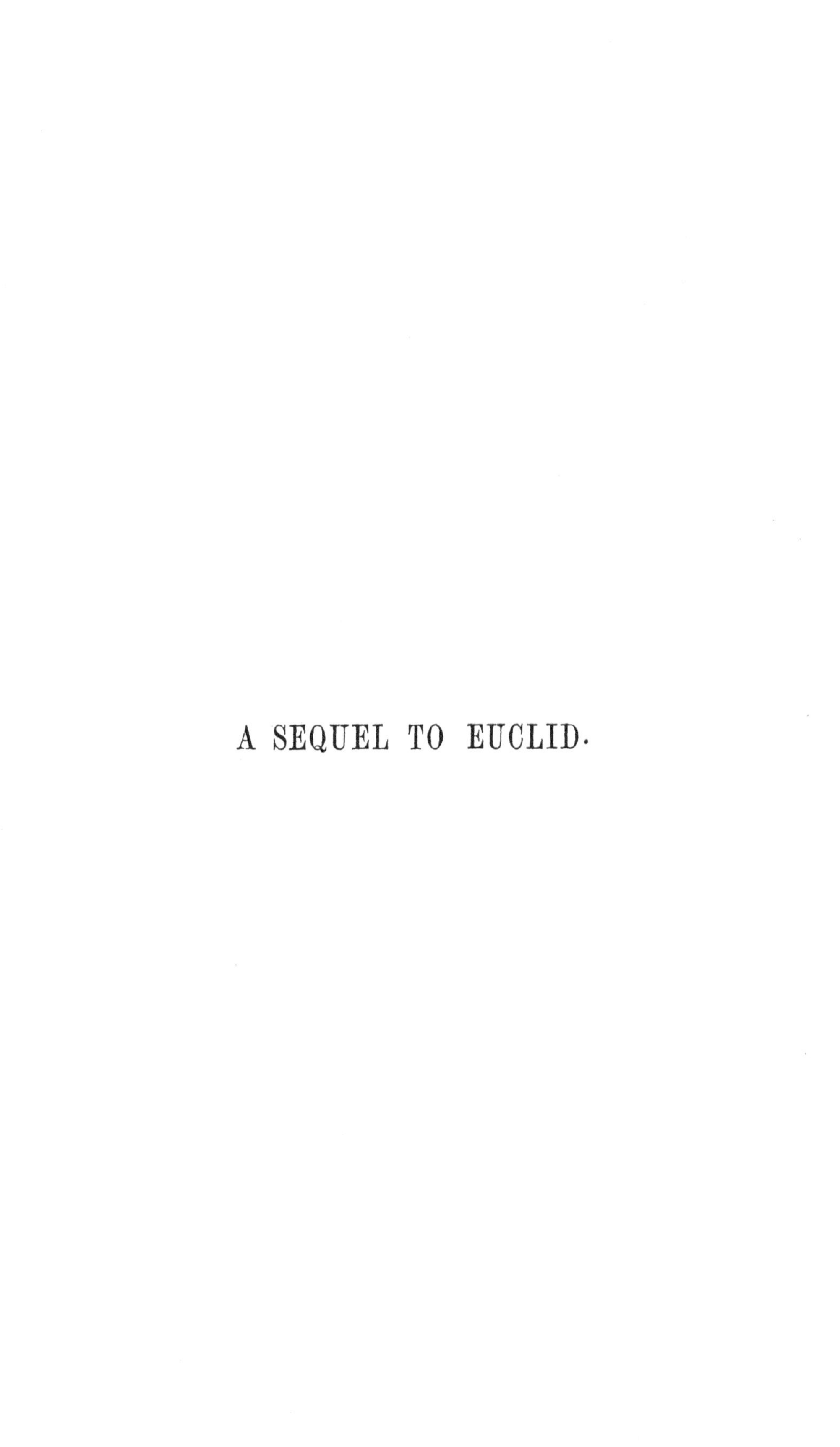

A SEQUEL TO EUCLID.

DUBLIN UNIVERSITY PRESS SERIES.

A SEQUEL TO THE FIRST SIX BOOKS

OF THE

ELEMENTS OF EUCLID,

CONTAINING

AN EASY INTRODUCTION TO MODERN GEOMETRY,

With Numerous Examples,

BY

JOHN CASEY, LL.D., F.R.S.,

Fellow of the Royal University of Ireland;
Member of the Council of the Royal Irish Academy;
Member of the Mathematical Societies of London and France;
Corresponding Member of the Royal Society of Sciences of Liege; and
Professor of the Higher Mathematics and Mathematical Physics
in the Catholic University of Ireland.

FIFTH EDITION, REVISED AND ENLARGED.

DUBLIN: HODGES, FIGGIS, & CO., GRAFTON-ST.
LONDON: LONGMANS, GREEN, & CO.
1888.

DUBLIN:
PRINTED AT THE UNIVERSITY PRESS,
BY PONSONBY AND WELDRICK.

PREFACE.

I HAVE endeavoured in this Manual to collect and arrange all those Elementary Geometrical Propositions not given in Euclid which a Student will require in his Mathematical Course. The necessity for such a Work will be obvious to every person engaged in Mathematical Tuition. I have been frequently obliged, when teaching the Higher Mathematics, to interrupt my demonstrations, in order to prove some elementary Propositions on which they depended, but which were not given in any book to which I could refer. The object of the present little Treatise is to supply that want.

The following is the plan of the Work. It is divided into five Chapters, corresponding to Books I., II., III., IV., VI. of Euclid. The Supplements to Books I.–IV. consist of two Sections each, namely, Section I., Additional Propositions; Section II., Exercises. This part will be found to contain original proofs of some of the most elegant Propositions in Geometry. The Supplement to Book VI. is the most important; it embraces more than half the work, and consists of eight Sections, as follows:—I., Additional Propositions; II., Centres of Similitude; III., Theory of Harmonic Section; IV., Theory of Inversion; V., Coaxal Circles; VI., Theory of Anharmonic Section; VII., Theory of Poles and Polars, and Reciprocation; VIII., Miscellaneous Exercises. Some of the Propositions in these Sections have first appeared in Papers published by myself; but the greater number have been selected from the writings of CHASLES, SALMON, and TOWNSEND. For the proofs given by these authors, in some instances others have been substituted, but in no case except where by doing so they could be made more simple and elementary.

The present edition is greatly enlarged: the new matter, consisting of recent discoveries in Geometry, is contained in a Supplemental Chapter. Several of the Demonstrations, and some of the Propositions in this Chapter, are original, in particular the Theory of Harmonic Polygons, in Section VI. A large number of the Miscellaneous Exercises are also original.

In collecting and arranging these additions I have received valuable assistance from Professor NEUBERG, of the University of Liege, and from M. BROCARD (after whom the Brocard Circle is named). The other writers to whom I am indebted are mentioned in the text.

The principles of Modern Geometry contained in the Work are, in the present state of science, indispensable in Pure and Applied Mathematics,* and in Mathematical Physics;† and it is important that the Student should become early acquainted with them.

JOHN CASEY.

86, South Circular Road,
Dublin, *Aug.* 31, 1886.

---

* See Chalmers' "Graphical Determination of Forces in Engineering Structures," and Lévy's "Statique Graphique."

† See Sir W. Thomson's Papers on "Electrostatics and Magnetism"; Clerk Maxwell's "Electricity."

The following selected Course is recommended to Junior Students :—

Book I.—Additional Propositions,			1–22, inclusive.
Book II.—	,,	,,	1–12, inclusive.
Book III.—	,,	,,	1–28, inclusive.
Book IV.—	,,	,,	1–9, omitting 6, 7.
Book VI.—	,,	,,	1–12, omitting 6, 7.

Book VI.—Sections II., III., IV., V., VI., VII., omitting Proof of Feuerbach's Theorem, page 105 ; Prop. XIV., page 109 ; Second Proof of Prop. XVI., page 112 ; Second Solution of Prop. X., p. 123.

CONTENTS.

.he articles marked with asterisks may be omitted on a first reading.

BOOK I.

SECTION I.

SECTION II.

BOOK II.

SECTION I.

SECTION II.

BOOK III.

SECTION I.

SECTION II.

BOOK IV.

SECTION I.

SECTION II.

BOOK VI.

SECTION I.

SECTION II.

SECTION III.

SECTION IV.

SECTION V.

SECTION VI.

SECTION VII.

SECTION VIII.

SUPPLEMENTARY CHAPTER.

SECTION I.

SECTION II.

SECTION III.

SECTION IV.

SECTION V.

SECTION VI.

SECTION VII.

ERRATA.

Page 158, line 22, *for* equal *read* parallel.

,, 183, ,, 8, *for* circles *read* circle.

,, 226, ,, 12, *for* passing *read* passing through.

ERRATA.

Page 168, line 7, *for* isotomic *read* isotomic conjugates.
,, 187, ,, 2, *after* centres *supply* are.
,, 209, lines 3, 10, 11, *for* $\delta^2 R^2$ *read* δ^2 / R^2.
,, 209, line 25, *after* $A_2 A_{n-2}$ *supply* &c.
,, 216, ,, 6, *for* radius *read* diameter.
,, 248, ,, 15, *for* 2 : 1 *read* ratio 2 : 1.

BOOK FIRST.

SECTION I.

Additional Propositions.

In the following pages the Propositions of the text of Euclid will be referred to by Roman numerals enclosed in brackets, and those of the work itself by the Arabic. The number of the book will be given only when different from that under which the reference occurs.

For the purpose of saving space, the following symbols will be employed:—

Circle will be denoted by		⊙
Triangle	,,	△
Parallelogram	,,	▭
Parallel	,,	∥
Angle	,,	∠
Perpendicular	,,	⊥

In addition to the foregoing, we shall employ the usual symbols of Algebra, and other contractions whose meanings will be so obvious as not to require explanation.

Prop. 1.—*The diagonals of a parallelogram bisect each other.*

Let ABCD be the ▭, its diagonals AC, BD bisect each other.

Dem.—Because AC meets the ||s AB, CD, the ∠ BAO = DCO. In like manner, the ∠ ABO = CDO (xxix.), and the side AB = side CD (xxxiv.); ∴ AO = OC; BO = OD (xxvi.)

Cor. 1.—If the diagonals of a quadrilateral bisect each other it is a ▭.

Cor. 2.—If the diagonals of a quadrilateral divide it into four equal triangles, it is a ▭.

Prop. 2.—*The line* DE *drawn through the middle point* D *of the side* AB *of a triangle, parallel to a second side* BC, *bisects the third side* AC.

Dem.—Through C draw CF || to AB, meeting DE produced in F. Since BCFD is a ▭, CF = BD (xxxiv.); but BD = AD (hyp.); ∴ CF = AD.

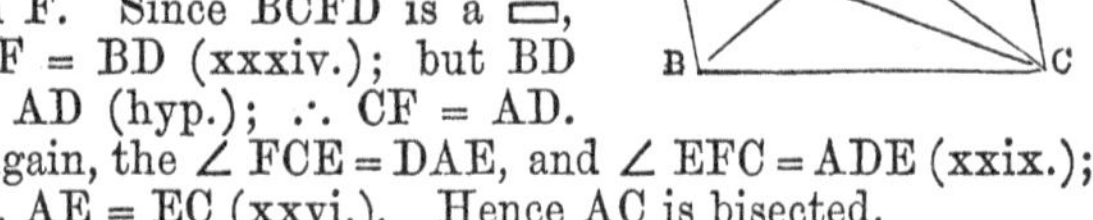

Again, the ∠ FCE = DAE, and ∠ EFC = ADE (xxix.); ∴ AE = EC (xxvi.). Hence AC is bisected.

Cor.—DE = $\frac{1}{2}$ BC. For DE = EF = $\frac{1}{2}$ DF.

Prop. 3.—*The line* DE *which joins the middle points* D *and* E *of the sides* AB, AC *of a triangle is parallel to the base* BC.

Dem.—Join BE, CD (fig. 2), then △ BDE = ADE (xxxviii.), and CDE = ADE; therefore the △ BDE = CDE, and the line DE is || to BC (xxxix.).

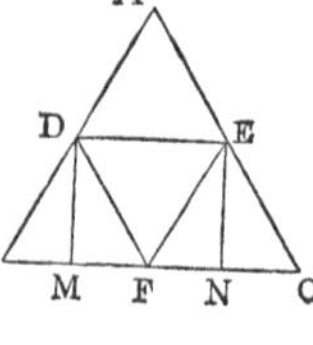

Cor. 1.—If D, E, F be the middle points of the sides AB, AC, BC of a △, the four △s into which the lines DE, EF, FD divide the △ABC are all equal. This follows from (xxxiv.), because the figures ADFE, CEDF, BFED, are ▭s.

Cor. 2.—If through the points D, E, any two ∥s be drawn meeting the base BC in two points M, N, the ▭ DENM is = $\frac{1}{2}$ △ ABC. For DENM = ▭ DEFB (xxxv.).

Def.—*When three or more lines pass through the same point they are said to be concurrent.*

Prop. 4.—*The bisectors of the three sides of a triangle are concurrent.*

Let BE, CD, the bisectors of AC, AB, intersect in O; the Prop. will be proved by showing that AO produced bisects BC. Through B draw BG ∥ to CD, meeting AO produced in G; join CG. Then, because DO bisects AB, and is ∥ to BG, it bisects AG (2) in O. Again, because OE bisects the sides AG, AC, of the △ AGC, it is ∥ to GC (3). Hence the figure OBGC is a ▭, and the diagonals bisect each other (1); ∴ BC is bisected in F.

Cor.—The bisectors of the sides of a △ divide each other in the ratio of 2 : 1.

Because AO = OG and OG = 2OF, AO = 2OF.

Prop. 5.—*The middle points* E, F, G, H *of the sides* AC, BC, AD, BD *of two triangles* ABC, ABD, *on the same base* AB, *are the angular points of a parallelogram, whose area is equal to half sum or half difference of the areas of the triangles, according as they are on opposite sides, or on the same side of the common base.*

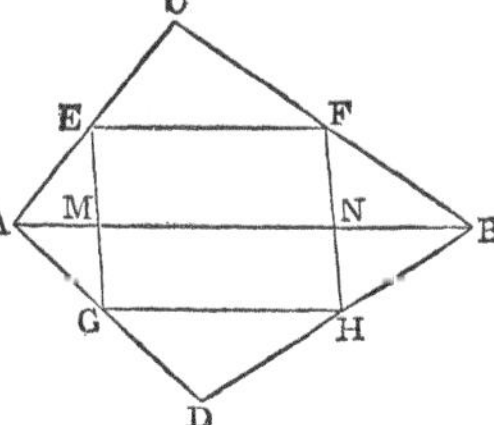

Dem. 1. Let the △s be on opposite sides. The

figure EFHG is evidently a ▭, since the opposite sides EF, GH are each ∥ to AB (3), and = ½ AB (Prop. 2, *Cor.*). Again, let the lines EG, FH meet AB in the points M, N; then ▭ EFNM = ½ △ ABC (Prop. 3, *Cor.* 2), and ▭ GHNM = ½ △ ABD. Hence ▭ EFHG = ½ (ABC + ABD).

Dem. 2.—When ABC, ABD are on the same side of AB, we have evidently ▭ EFGH = EFNM − GHNM = ½ (ABC − ABD).

Observation.—The second case of this proposition may be inferred from the first if we make the convention of regarding the sign of the area of the △ ABD to change from positive to negative, when the △ goes to the other side of the base. This affords a simple instance of a convention universally adopted by modern geometers, namely—when a geometrical magnitude of any kind, which varies continuously according to any law, passes through a zero value to give it the algebraic signs, plus and minus, on different sides of the zero—in other words, to suppose it to change sign in passing through zero, unless zero is a maximum or minimum.

Prop. 6.—*If two equal triangles* ABC, ABD *be on the same base* AB, *but on opposite sides, the line joining the vertices* C, D *is bisected by* AB.

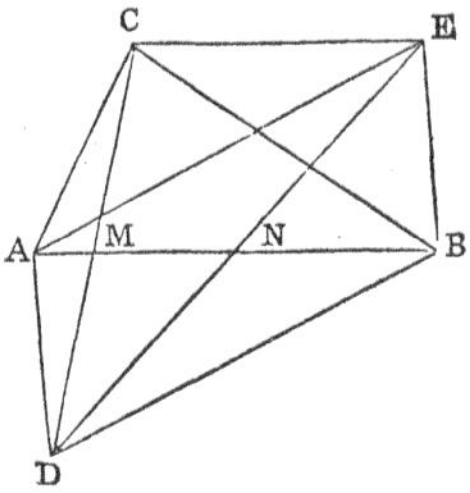

Dem.—Through A and B draw AE, BE ∥ respectively to BD, AD; join EC. Now, since AEBD is a ▭, the △ AEB = ADB (xxxiv.); but ADB = ACB (hyp.); ∴ AEB = ACB; ∴ CE is ∥ to AB (xxxix.). Let CD, ED meet AB in the points M, N, respectively. Now, since AEBD is a ▭, ED is bisected in N (1); and since NM is ∥ to EC, CD is bisected in M (2).

Cor.—If the line joining the vertices of two △s on the same base, but on opposite sides, be bisected by the base, the △s are equal.

Prop. 7.—*If the opposite sides* AB, CD *of a quadrilateral* *meet in* P, *and if* G, H *be the middle points of the diagonals* AC, BD, the *triangle* PGH = $\frac{1}{4}$ *the quadrilateral* ABCD.

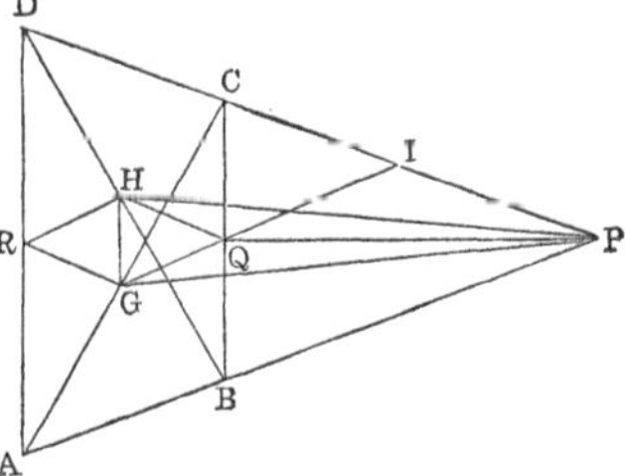

Dem.—Bisect the sides BC, AD in Q and R; join QH, QG, QP, RH, RG. Now, since QG is ∥ to AB (3), if produced it will bisect PC; then, since CP, joining the vertices of the △s CGQ, PGQ on the same base GQ, but on opposite sides, is bisected by GQ produced, the △ PGQ = CGQ (Prop. 6, *Cor.*) = $\frac{1}{4}$ ABC.

In like manner PHQ = $\frac{1}{4}$ BCD. Again, the ▭ GQHR = $\frac{1}{2}$ (ABD − ABC) (5); ∴ △ QGH = $\frac{1}{4}$ ABD − $\frac{1}{4}$ ABC: hence, △ PGH = $\frac{1}{4}$ (ABC + BCD + ABD − ABC) = $\frac{1}{4}$ quadrilateral ABCD.

Cor.—The middle points of the three diagonals of a complete quadrilateral are collinear (*i. e.* in the same right line). For, let AD and BC meet in S, then SP will be the third diagonal; join S and P to the middle points G, H of the diagonals AC, BD; then the △s SGH, PGH, being each = $\frac{1}{4}$ quadrilateral ABCD, are = to one another; ∴ GH produced bisects SP (6).

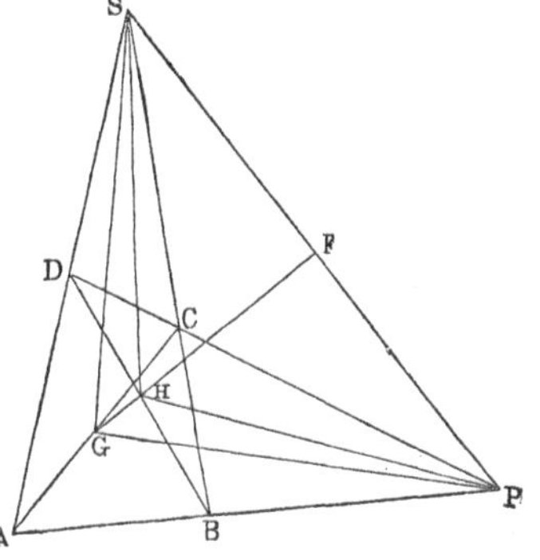

Def.—*If a variable point moves according to any law, the path which it describes is termed its locus.*

Thus, if a point P moves so as to be always at the same distance from a fixed point O, the locus of P is a ⊙, whose centre is O and radius = OP. Or, again, if

A and B be two fixed points, and if a variable point P moves so that the area of the △ ABP retains the same value during the motion, the locus of P will be a right line ∥ to AB.

Prop. 8.—*If* AB, CD *be two lines given in position and magnitude, and if a point* P *moves so that the sum of the areas of the triangles* ABP, CDP *is given, the locus of* P *is a right line.*

Dem.—Let AB, CD intersect in O; then cut off OE = AB, and OF = CD; join OP, EP, EF, FP; then △ APB = OPE, and CPD = OPF; hence the sum of the areas of the △s OEP, OFP is given; ∴ the area of the quadrilateral OEPF is given; but the △ OEF is evidently given; ∴ the area of the △ EFP is given, and the base EF is given; ∴ the locus of P is a right line ∥ to EF.

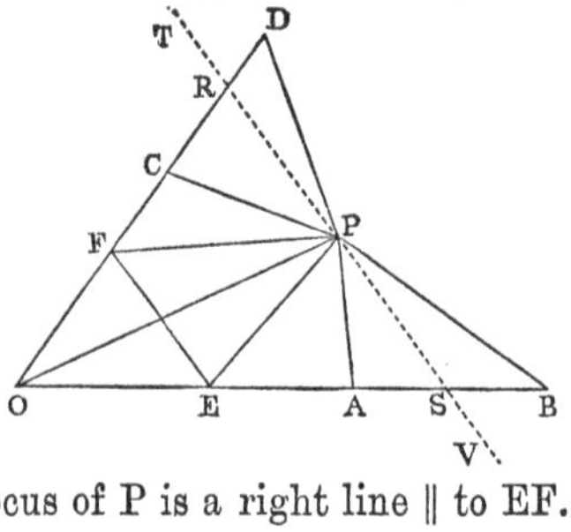

Let the locus in this question be the dotted line in the diagram. It is evident, when the point P coincides with R, the area of the △ CDP vanishes; and when the point P passes to the other side of CD, such as to the point T, the area of the △ CDP must be regarded as negative. Similar remarks hold for the △ APB and the line AB. This is an instance of the principle (see 5, note) that the area of a △ passes from positive to negative as compared with any given △ in its own plane, when (in the course of any continuous change) its *vertex crosses its base.*

Cor. 1.—If m and n be any two multiples, and if we make OE = mAB and OF = nCD, we shall in a similar way have the locus of the point P when m times △ ABP + n times CDP is given; viz., it will be a right line ∥ to EF.

Cor. 2.—If the line CD be produced through O, and if we take in the line produced, OF′ = nCD, we shall get the locus of P when m times △ ABP − n times CDP is given.

Cor. 3.—If three lines, or in general any number of

lines, be given in magnitude and position, and if m, n, p, q, &c., be any system of multiples, all positive, or some positive and some negative, and if the area of m times △ ABP + n times CDP + p times GHP + &c., be given, the locus of P is a right line.

Cor. 4.—If ABCD be a quadrilateral, and if P be a point, so that the sum of the areas of the △s ABP, CDP is half the area of the quadrilateral, the locus of P is a right line passing through the middle points of the three diagonals of the quadrilateral.

Prop. 9.—*To divide a given line* AB *into two parts, the difference of whose squares shall be equal to the square of a given line* CD.

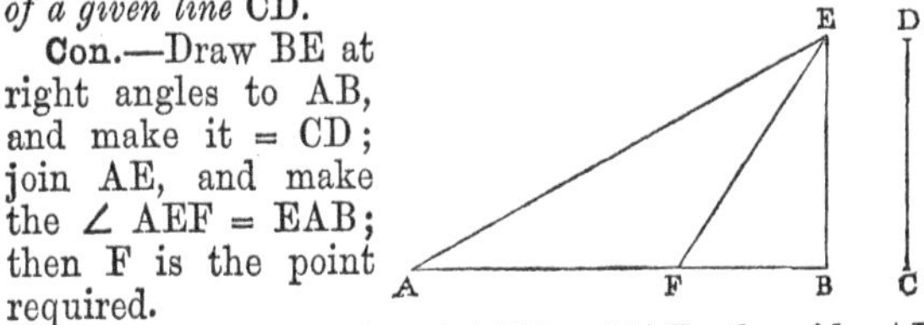

Con.—Draw BE at right angles to AB, and make it = CD; join AE, and make the ∠ AEF = EAB; then F is the point required.

Dem.—Because the ∠ AEF = EAF, the side AF = FE; ∴ $AF^2 = FE^2 = FB^2 + BE^2$; ∴ $AF^2 - FB^2 = BE^2$; but $BE^2 = CD^2$; ∴ $AF^2 - FB^2 = CD^2$.

If CD be greater than AB, BE will be greater than AB, and the ∠ EAB will be greater than the ∠ AEB; hence the line EF, which makes with AE the ∠ AEF = ∠ EAB, will fall at the other side of EB, and the point F will be in the line AB produced. The point F is in this case a point of external division.

Prop. 10.—*Given the base of a triangle in magnitude and position, and given also the difference of the squares of its sides, to find the locus of its vertex.*

Let ABC be the △ whose base AB is given; let fall the ⊥ CP on AB; then

$$AC^2 = AP^2 + CP^2; \qquad \text{(xlvii.)}$$
$$BC^2 = BP^2 + CP^2;$$

therefore $$AC^2 - BC^2 = AP^2 - BP^2;$$

but $AC^2 - BC^2$ is given; ∴ $AP^2 - BP^2$ is given. Hence AB is divided in P into two parts, the difference of whose squares is given; ∴ P is a given point (9), and the line CP is given in position; and since the point C must be always on the line CP, the locus of C is a right line ⊥ to the base.

Cor.—The three ⊥s of a △ are concurrent. Let the ⊥s from A and B on the opposite sides be AD and BE, and let O be the point of intersection of these ⊥s.

Now, $AC^2 - AB^2 = OC^2 - OB^2$; (10)
and $AB^2 - BC^2 = OA^2 - OC^2$;
therefore $AC^2 - BC^2 = OA^2 - OB^2$.

Hence the line CO produced will be ⊥ to AB.

Prop. 11.—*If perpendiculars* AE, BF *be drawn from the extremities* A, B *of the base of a triangle on the internal bisector of the vertical angle, the line joining the middle point* G *of the base to the foot of either perpendicular is equal to half the difference of the sides* AC, BC.

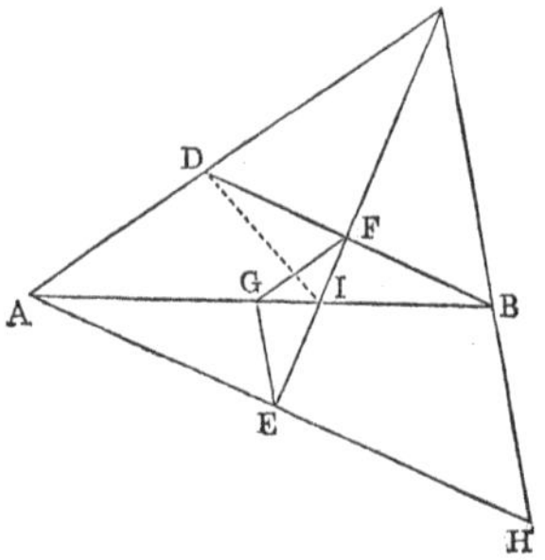

Dem.—Produce BF to D; then in the △s BCF, DCF there are evidently two ∠s and a side of one = respectively to two ∠s and a side of the other; ∴ CD = CB and FD = FB; hence AD is the difference of the sides AC, BC; and, since F and G are the middle points of the sides BD, BA; ∴ FG = ½ AD = ½ (AC − BC). In like manner EG = ½ (AC − BC).

Cor. 1.—By a similar method it may be proved that lines drawn from the middle point of the base to the feet of ⊥s from the extremities of the base on the bisector of the external vertical angle are each = half sum of AC and BC.

Cor. 2.—The ∠ ABD is = ½ difference of the base angles.

Cor. 3.—CBD is = half sum of the base angles.

Cor. 4.—The angle between CF and the ⊥ from C on AB = ½ difference of the base angles.

Cor. 5.—AID = difference of the base angles.

Cor. 6.—Given the base and the difference of the sides of a △, the locus of the feet of the ⊥s from the extremities of the base on the bisector of the internal vertical ∠ is a circle, whose centre is the middle point of the base, and whose radius = half difference of the sides.

Cor. 7.—Given the base of a △ and the sum of the sides, the locus of the feet of the ⊥s from the extremities of the base on the bisector of the external vertical ∠ is a circle, whose centre is the middle point of the base, and whose radius = half sum of the sides.

Prop. 12.—*The three perpendiculars to the sides of a triangle at their middle points are concurrent.*

Dem.—Let the middle points be D, E, F. Draw FG, EG ⊥ to AB, AC, and let these ⊥s meet in G; join GD: the prop. will be proved by showing that GD is ⊥ to BC. Join AG, BG, CG. Now, in the △s AFG and BFG, since AF = FB, and FG common, and the ∠ AFG = BFG, AG is = GB (iv.). In like manner AG = GC; hence BG = GC. And since the △s BDG, CDG have the side BD = DC and DG common, and the base BG = GC, the ∠ BDG = CDG (viii.); ∴ GD is ⊥ to BC.

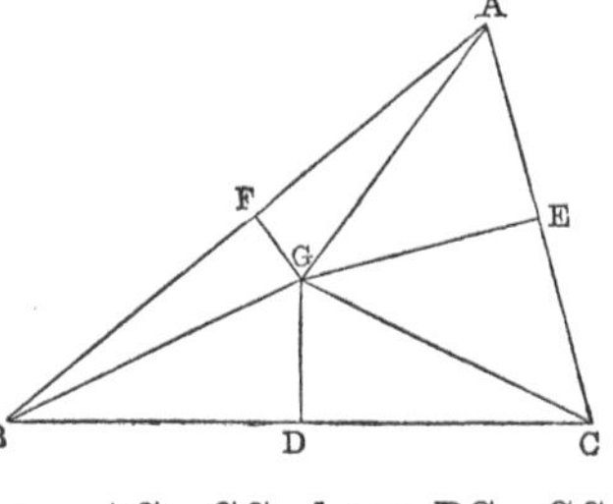

Cor. 1.—If the bisectors of the sides of the △ meet in H, and GH be joined and produced to meet any of the three ⊥s from the ∠s on the opposite sides; for instance, the ⊥ from A to BC, in the point I, suppose; then GH = $\frac{1}{2}$ HI. For DH = $\frac{1}{2}$ HA (*Cor.*, Prop. 4).

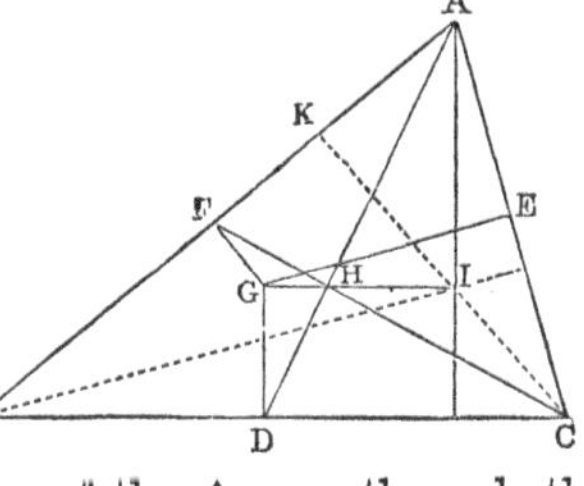

Cor. 2.—Hence the ⊥s of the △ pass through the point I. This is another proof that the ⊥s of a △ are concurrent.

Cor. 3.—The lines GD, GE, GF are respectively $= \frac{1}{2}$ IA, $\frac{1}{2}$ IB, $\frac{1}{2}$ IC.

Cor. 4.—The point of concurrence of ⊥s from the ∠s on the opposite sides, the point of concurrence of bisectors of sides, and the point of concurrence of ⊥s at middle points of sides of a △, are collinear.

Prop. 13.—*If two triangles* ABC, ABD, *be on the same base* AB *and between the same parallels, and if a parallel to* AB *intersect the lines* AC, BC, *in* E *and* F, *and the lines* AD, BD, *in* G *and* H, EF *is* = GH.

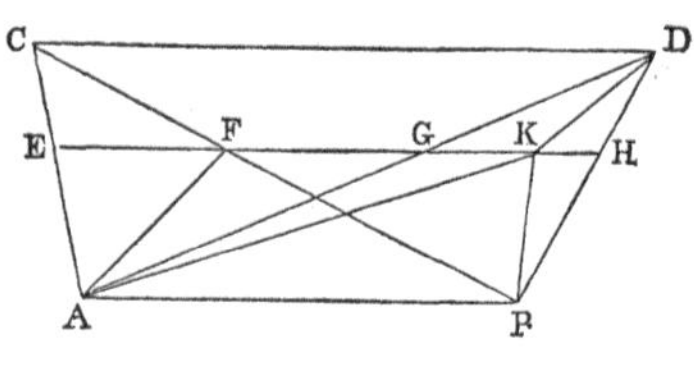

Dem.—If not, let GH be greater than EF, and cut off GK = EF. Join AK, KB, KD, AF; then (xxxviii.) △ AGK = AEF, and DGK = CEF, and (xxxvii.) ABK = ABF; ∴ the quadrilateral ABKD = △ ABC; but △ ABC = ABD; ∴ the quadrilateral ABKD = △ ABD, which is impossible. Hence EF = GH.

Cor. 1.—If instead of two △s on the same base and between the same ‖s, we have two △s on equal bases and between the same ‖s, the intercepts made by the sides of the △s on a ‖ to the line joining the vertices are equal.

Cor. 2.—The line drawn from the vertex of a △ to the middle point of the base bisects any line parallel to the base, and terminated by the sides of the triangle.

Prop. 14.—*To inscribe a square in a triangle.*

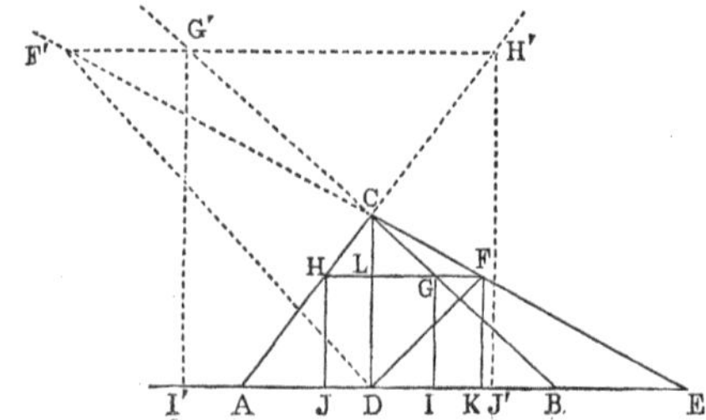

Con.—Let ABC be the △: let fall the ⊥ CD; cut off BE = AD; join EC; bisect the ∠ EDC by the line DF, meeting EC in F; through F draw a ∥ to AB, cutting the sides BC, AC in the points G, H; from G, H let fall the ⊥s GI, HJ: the figure GIJH is a square.

Dem.—Since the ∠ EDC is bisected by DF, and the ∠s K and L right angles, and DF common, FK = FL (xxvi.); but FL = GH (Prop. 13, *Cor.* 1), and FK = GI (xxxiv.); ∴ GI = GH, and the figure IGHJ is a square, and it is inscribed in the triangle.

Cor.—If we bisect the ∠ ADC by the line DF′, meeting EC produced in F′, and through F′ draw a line ∥ to AB meeting BC, and AC produced in G′, H′, and from G′, H′ let fall ⊥s G′I′, H′J′ on AB, we shall have an escribed square.

Prop. 15.—*To divide a given line* AB *into any number of equal parts.*

Con.—Draw through A any line AF, making an ∠ with AB; in AF take any point C, and cut off CD, DE, EF, &c., each = AC, until we have as many equal parts as the number into which we want to divide AB—say, for instance, four equal parts. Join BF; and draw CG, DH, EI, each ∥ to BF; then AB is divided into four equal parts.

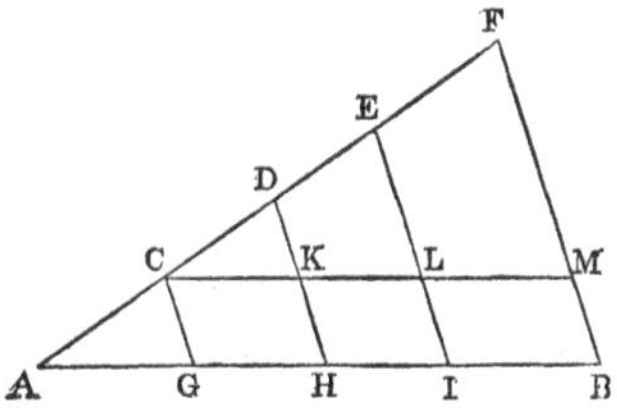

Dem.—Since ADH is a △, and AD is bisected in C, and CG is ∥ to DH; then (2) AH is bisected in G; ∴ AG = GH. Again, through C draw a line ∥ to AB, cutting DH and EI in K and L; then, since CD = DE, we have (2) CK = KL; but CK = GH, and KL = HI; ∴ GH = HI. In like manner, HI = IB. Hence the parts into which AB is divided are all equal.

This Proposition may be enunciated as a theorem as follows:—If one side of a △ be divided into any number of equal parts, and through the points of division lines be drawn ∥ to the base, these ∥s will divide the second side into the same number of equal parts.

Prop. 16.—*If a line* AB *be divided into* $(m + n)$ *equal parts, and suppose* AC *contains* m *of these parts, and* CB *contains* n *of them. Then, if from the points* A, C, B *perpendiculars* AD, CF, BE *be let fall on any line, then* mBE + nAD = $(m + n)$ CF.

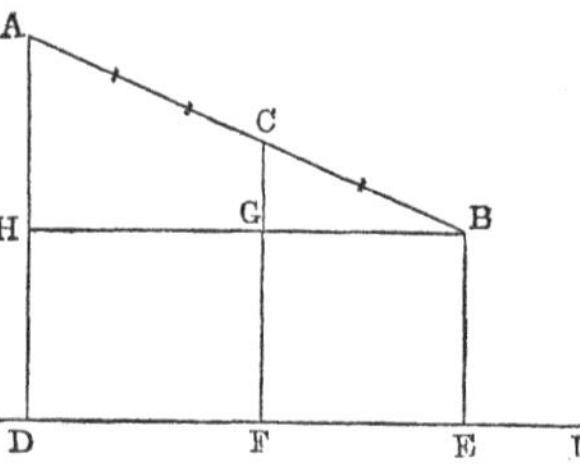

Dem. — Draw BH ‖ to ED, and through the points of division of AB imagine lines drawn ‖ to BH; these lines will divide AH into $m + n$ equal parts, and CG into n equal parts; ∴ n times AH=$(m+n)$ times CG; and since DH and BE are each = GF, we have n times HD + m times BE = $(m+n)$ times GF. Hence, by addition, n times AD + m times BE = $(m+n)$ times CF.

Def.—*The centre of mean position of any number of points* A, B, C, D, *&c., is a point which may be found as follows:—Bisect the line joining any two points* AB *in* G, *join* G *to a third point* C, *and divide* GC *in* H, *so that* GH = $\frac{1}{3}$ GC; *join* H *to a fourth point* D, *and divide* HD *in* K, *so that* HK = $\frac{1}{4}$HD, *and so on: the last point found will be the centre of mean position of the system of points.*

Prop. 17.—*If there be any system of points* A, B, C, D, *whose number is* n, *and if perpendiculars be let fall from these points on any line* L, *the sum of the perpendiculars from all the points on* L *is equal* n *times the perpendicular from the centre of mean position.*

Dem.—Let the ⊥s be denoted by AL, BL, CL, &c. Then, since AB is bisected in G, we have (16)

$$AL + BL = 2GL;$$

and since GC is divided into $(1 + 2)$ equal parts in H, so that HG contains one part and HC two parts; then 2GL + CL = 3HL;

$$\therefore AL + BL + CL = 3HL, \text{ \&c., \&c.}$$

Hence the Proposition is proved.

Cor.—If from any number of points ⊥s be let fall on any line passing through their mean centre, the sum of the ⊥s is zero. Hence some of the ⊥s must be negative, and we have the sum of the ⊥s on the positive side equal to the sum of those on the negative side.

Prop. 18.—*We may extend the foregoing Definition as follows:—Let there be any system of points* A, B, C, D, *&c., and a corresponding system of multiples a, b, c, d, &c., connected with them; then divide the line joining the points* AB *into* $(a+b)$ *equal parts, and let* AG *contain b of these parts, and* GB *contain a parts. Again, join* G *to a third point* C, *and divide* GC *into* $(a+b+c)$ *equal parts, and let* GH *contain c of these parts, and* HC *the remaining parts, and so on; then the point last found will be the mean centre for the system of multiples a, b, c, d, &c.*

From this Definition we may prove exactly the same as in Prop. 17, that if AL, BL, CL, &c., be the ⊥s from the points A, B, C, &c., on any line L, then

$$a \,.\, AL + b \,.\, BL + c \,.\, CL + d \,.\, DL + \&c.$$

$=(a+b+c+d+\&c.)$ times the ⊥ from the centre of mean position on the line L.

Def.—*If a geometrical magnitude varies its position continuously according to any law, and if it retains the same value throughout, it is said to be a constant; but if it goes on increasing for some time, and then begins to decrease, it is said to be a maximum at the end of the increase: again, if it decreases for some time, and then begins to increase, it is a minimum when it commences to increase.*

From these Definitions it will be seen that a maximum value is greater than the ones which immediately precede and follow; and that a minimum is less than the value of that which immediately precedes, and less than that which immediately follows. We give here a few simple but important Propositions bearing on this part of Geometry.

Prop. 19.—*Through a given point* P *to draw a line which shall form, with two given lines* CA, CB, *a triangle of minimum area.*

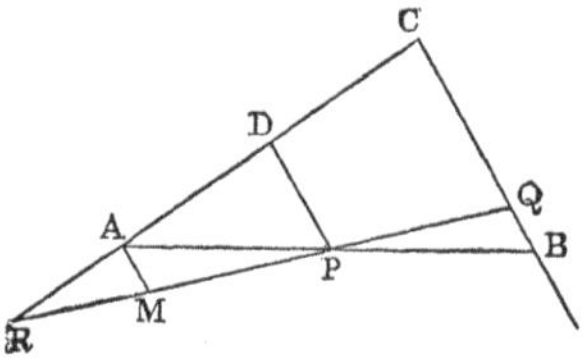

Con.—Through P draw PD ∥ to CB; cut off AD = CD; join AP, and produce to B. Then AB is the line required.

Dem.—Let RQ be any other line through P; draw AM ∥ to CB. Now, because AD = DC, we have AP = PB; and the △s APM and QPB have the ∠s APM, AMP respectively equal to BPQ, BQP, and the sides AP and PB equal to one another; ∴ the triangles are equal; hence the △ APR is greater than BPQ: to each add the quadrilateral CAPQ, and we get the △ CQR greater than ABC.

Cor. 1.—The line through the point P which cuts off the minimum triangle is bisected in that point.

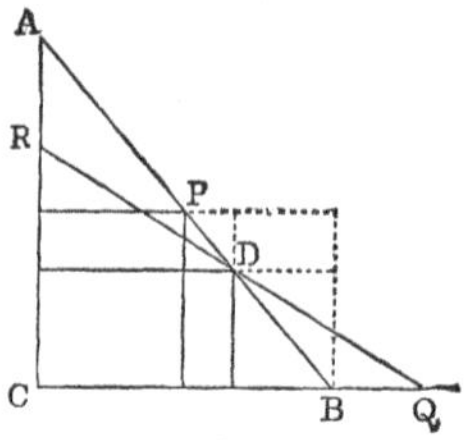

Cor. 2.—If through the middle point P, and through any other point D of the side AB of the △ ABC we draw lines ∥ to the remaining sides, so as to form two inscribed ▭s CP, CD, then CP is greater than CD.

Dem.—Through D draw QR, so as to be bisected in D; then the △ ABC is greater than CQR; but the ▭s are halves of the △s; hence CP is greater than CD.

A very simple proof of this *Cor.* can also be given by means of (xliii.)

Prop. 20.—*When two sides of a triangle are given in magnitude, the area is a maximum when they contain a right angle.*

Dem.—Let BAC be a △ having the ∠ A right; with A as centre and AC as radius, describe a ⊙; take any other point D in the circumference; it is evident the Prop. will be proved by showing that the △ BAC is greater than BAD. Let fall the ⊥ DE; then (xix.) AD is greater than DE; ∴ AC is greater than DE; and since the base AB is common, the △ ABC is greater than ABD.

Cor.—If the diagonals of a quadrilateral be given in magnitude, the area is a maximum when they are at right angles to each other.

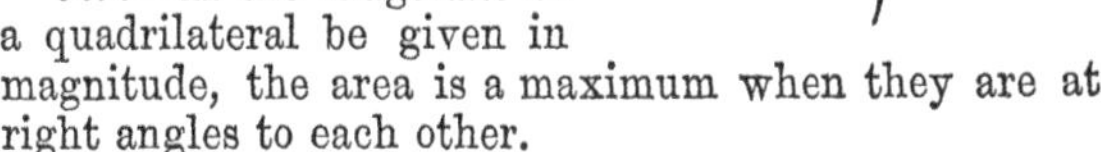

Prop. 21.—*Given two points,* A, B: *it is required to find a point* P *in a given line* L, *so that* AP + PB *may be a minimum.*

Con.—From B let fall the ⊥ BC on L; produce BC to D, and make CD = CB; join AD, cutting L in P; then P is the point required.

Dem.—Join PB, and take any other point Q in L; join AQ, QB, QD. Now, since BC = CD and CP common, and the ∠s at C right ∠s, we have BP = PD. In like manner BQ = QD; to these equals add respectively AP and AQ, and we have AD = AP + PB, and AQ + QD = AQ + QB; but AQ + QD is greater than AD; ∴ AQ + QB is greater than AP + PB.

Cor. 1.—The lines AP, PB, whose sum is a minimum, make equal angles with the line L.

Cor. 2.—The perimeter of a variable △, inscribed in a fixed △, is a minimum when the sides of the former make equal ∠s with the sides of the latter. For, suppose one side of the inscribed △ to remain fixed while the two remaining sides vary, the sum of the varying sides will be a minimum when they make equal ∠s with the side of the fixed triangle.

Cor. 3.—Of all polygons whose vertices lie on fixed lines, that of minimum perimeter is the one whose several angles are bisected externally by the lines on which they move.

Prop. 22.—*Of all triangles having the same base and area, the perimeter of an isosceles triangle is a minimum.*

Dem.—Since the △s are all equal in area, the vertices must lie on a line ∥ to the base, and the sides of an isosceles △ will evidently make equal ∠s with this parallel; hence their sum is a minimum.

Cor.—Of all polygons having the same number of sides and equal areas, the perimeter of an equilateral polygon is a minimum.

Prop. 23.—*A large number of deducibles may be given in connexion with Euclid,* fig., Prop. xlvii. *We insert a few here, confining ourselves to those that may be proved by the First Book.*

(1). The transverse lines AE, BK are ⊥ to each other. For, in the △s ACE, BCK, which are in every respect equal, the ∠ EAC = BKC, and the ∠ AQO = KQC; hence the angle AOQ = KCQ, and is ∴ a right angle.

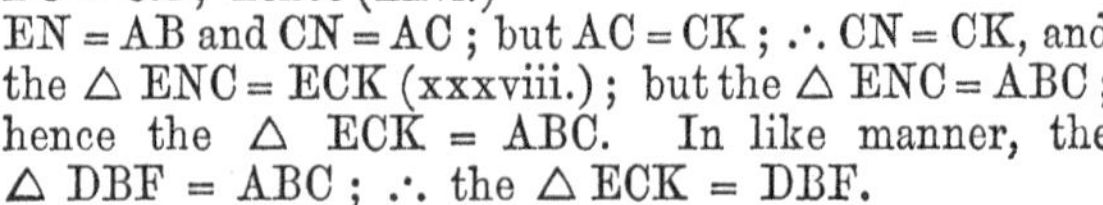

(2). △ KCE = DBF.

Dem.—Produce KC, and let fall the ⊥ EN. Now, the ∠ ACN = BCE, each being a right angle; ∴ the ∠ ACB = ECN, and ∠BAC = ENC, each being a right angle, and side BC = CE; hence (xxvi.) EN = AB and CN = AC; but AC = CK; ∴ CN = CK, and the △ ENC = ECK (xxxviii.); but the △ ENC = ABC; hence the △ ECK = ABC. In like manner, the △ DBF = ABC; ∴ the △ ECK = DBF.

(3). $EK^2 + FD^2 = 5BC^2$.

Dem.—$EK^2 = EN^2 + NK^2$ (xlvii.);

but $\quad EN = AB$, and $NK = 2AC$;
therefore $\quad EK^2 = AB^2 + 4AC^2$.

In like manner

$$FD^2 = 4AB^2 + AC^2;$$

therefore $\quad EK^2 + FD^2 = 5(AB^2 + AC^2) = 5BC^2$.

(4). The intercepts AQ, AR are equal.

(5). The lines CF, BK, AL are concurrent.

SECTION II.

Exercises.

1. The line which bisects the vertical ∠ of an isosceles △ bisects the base perpendicularly.

2. The diagonals of a quadrilateral whose four sides are equal bisect each other perpendicularly.

3. If the line which bisects the vertical ∠ of a △ also bisects the base, the △ is isosceles.

4. From a given point in one of the sides of a △ draw a right line bisecting the area of the △.

5. The sum of the ⊥s from any point in the base of an isosceles △ on the equal sides is = to the ⊥ from one of the base angles on the opposite side.

6. If the point be taken in the base produced, prove that the difference of the ⊥s on the equal sides is = to the ⊥ from one of the base angles on the opposite side; and show that, having regard to the convention respecting the signs plus and minus, this theorem is a case of the last.

7. If the base BC of a △ be produced to D, the ∠ between the bisectors of the ∠s ABC, ACD = half ∠ BAC.

8. The bisectors of the three internal angles of a △ are concurrent.

9. Any two external bisectors and the third internal bisector of the angles of a △ are concurrent.

10. The quadrilaterals formed either by the four external or the four internal bisectors of the angles of any quadrilateral have their opposite ∠s = two right ∠s.

11. Draw a right line ∥ to the base of a △, so that

(1). Sum of lower segments of sides shall be = to a given line.
(2). Their difference shall be = to a given line.
(3). The ∥ shall be = sum of lower segments.
(4). The ∥ shall be = difference of lower segments.

12. If two lines be respectively ⊥ to two others, the ∠ between the former is = to the ∠ between the latter.

13. If two lines be respectively ∥ to two others, the ∠ between the former is = to the ∠ between the latter.

14. The △ formed by the three bisectors of the external angles of a △ is such that the lines joining its vertices to the ∠s of the original △ will be its ⊥s.

15. From two points on opposite sides of a given line it is required to draw two lines to a point in the line, so that their difference will be a maximum.

16. State the converse of Prop. xvii.

17. Give a direct proof of Prop. xix.

18. Given the lengths of the bisectors of the three sides of a △: construct it.

19. If from any point ⊥s be drawn to the three sides of a △, prove that the sum of the squares of three alternate segments of the sides = the sum of squares of the three remaining segments.

20. Prove the following theorem, and infer from it Prop. xlvii.: If CQ, CP be ▱s described on the sides CA, CB of a △, and if the sides ∥ to CA, CB be produced to meet in R, and RC joined, a ▱ described on AB with sides = and ∥ to RC shall be = to the sum of the ▱s CQ, CP.

21. If a square be inscribed in a △, the rectangle under its side and the sum of base and altitude = twice the area of the △.

22. If a square be escribed to a △, the rectangle under its side and the difference of the base and altitude = twice the area of the △.

23. Given the difference between the diagonal and side of a square: construct it.

24. The sum of the squares of lines joining any point in the plane of a rectangle to one pair of opposite angular points = sum of the squares of the lines drawn to the two remaining angular points.

25. If two lines be given in position, the locus of a point equidistant from them is a right line.

26. In the same case the locus of a point, the sum or the difference of whose distances from them is given, is a right line.

27. Given the sum of the perimeter and altitude of an equilateral △: construct it.

28. Given the sum of the diagonal and two sides of a square: construct it.

29. From the extremities of the base ∠s of an isosceles △ right lines are drawn ⊥ to the sides, prove that the base ∠s of the △ are each = half the ∠ between the ⊥s.

30. The line joining the middle point of the hypotenuse of a right-angled triangle to the right angle is = half the hypotenuse.

31. The lines joining the feet of the ⊥s of a △ form an inscribed △ whose perimeter is a minimum.

32. If from the extremities A, B of the base of a △ ABC ⊥s AD, BE be drawn to the opposite sides, prove that

$$AB^2 = AC \cdot AE + BC \cdot BD.$$

33. If A, B, C, D, &c., be any number (n) of points on a line, and O their centre of mean position; then, if P be any other point on the line,

$$AP + BP + CP + DP + \&c. = nOP.$$

34. If O, O′ be the centres of mean position for two systems of collinear points, A, B, C, D, &c., A,′ B,′ C,′ D,′ &c., each system having the same number (n) of points; then

$$nOO' = AA' + BB' + CC' + DD' + \&c.$$

35. If G be the point of intersection of the bisectors of the ∠s A, B of a △, right-angled at C, and GD a ⊥ on AB; then, the rectangle AD . DB = area of the △.

36. The sides AB, AC of a △ are bisected in D, E; CD, BE intersect in F: prove △ BFC = quadrilateral ADFE.

37. If lines be drawn from a fixed point to all the points of the circumference of a given ⊙, the locus of all their points of bisection is a ⊙.

38. Show by drawing ∥ lines how to construct a △ = to any given rectilineal figure.

39. ABCD is a ▭: show that if B be joined to the middle point of CD, and D to the middle point of AB, the joining lines will trisect AC.

40. The equilateral △ described on the hypotenuse of a right-angled △ = sum of equilateral △s described on the sides.

41. If squares be described on the sides of any △, and the adjacent corners joined, the three △s thus formed are equal.

42. If AB, CD be opposite sides of a ▭, P any point in its plane; then △ PBC = sum or difference of the △s CDP, ACP, according as P is outside or between the lines AC and BD.

43. If equilateral △s be described on the sides of a right-angled △ whose hypotenuse is given in magnitude and position, the locus of the middle point of the line joining their vertices is a ⊙.

44. If CD be a ⊥ on the base AB of a right-angled △ ABC, and if E, F be the centres of the ⊙s inscribed in the △s ACD, BCD, and if EG, FH be lines through E and F ∥ to CD, meeting AC, BC in G, H; then CG = CH.

45. If A, B, C, D, &c., be any system of collinear points, O their mean centre for the system of multiples a, b, c, d, &c.; then, if P be any other point in the line,

$$(a+b+c+d+\&c.)\,\mathrm{OP} = a\,.\,\mathrm{AP} + b\,.\,\mathrm{BP} + c\,.\,\mathrm{CP} + d\,.\,\mathrm{DP} + \&c.$$

46. If O, O′ be the mean centres of the two systems of points A, B, C, D, &c., A′, B′, C′, D′, &c., on the same line L, for a common system of multiples a, b, c, d, &c.; then

$$(a+b+c+d+\&c.)\,\mathrm{OO'} = a\,.\,\mathrm{AA'} + b\,.\,\mathrm{BB'} + c\,.\,\mathrm{CC'} + d\,.\,\mathrm{OO'} + \&c.$$

BOOK SECOND.

SECTION I.

Additional Propositions.

Prop. 1.—*If* ABC *be an isosceles triangle, whose equal sides are* AC, BC; *and if* CD *be a line drawn from* C *to any point* D *in the base* AB; *then* $AD \cdot DB = BC^2 - CD^2$.

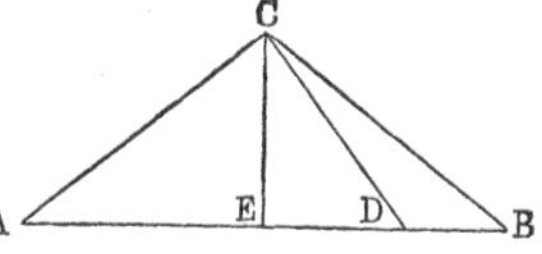

Dem.—Let fall the ⊥ CE; then AB is bisected in E and divided unequally in D;

therefore	$AD \cdot DB + ED^2 = EB^2$;
adding to each side	EC^2;
therefore	$AD \cdot DB + CD^2 = BC^2$; (I. xlvii.)
therefore	$AD \cdot DB = BC^2 - CD^2$.

Cor.—If the point be in the base produced, we shall have $AD \cdot BD = CD^2 - CB^2$. If we consider that DB changes its sign when D passes through B, we see that this case is included in the last.

Prop. 2.—*If* ABC *be any triangle,* D *the middle point of* AB, *then* $AC^2 + BC^2 = 2AD^2 + 2DC^2$.

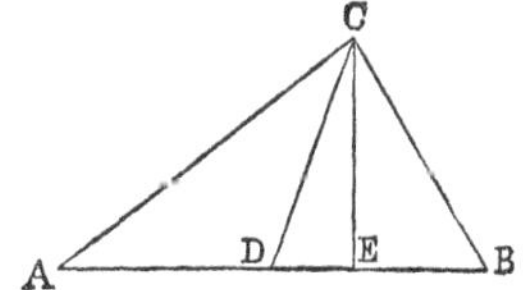

Dem.—Let fall the ⊥ EC.

$$AC^2 = AD^2 + DC^2 + 2AD \cdot DE; \qquad \text{(xii.)}$$
$$BC^2 = BD^2 + DC^2 - 2DB \cdot DE. \qquad \text{(xiii.)}$$

Hence, by addition, since AD = DB,

$$AC^2 + BC^2 = 2AD^2 + 2DC^2.$$

This is a simple case of a very general Prop., which we shall prove, on the properties of the centre of mean position for any system of points and any system of multiples. The Props. ix. and x. of the Second Book are particular cases of this Prop., viz., when the point C is in the line AB or the line AB produced.

Cor.—If the base of a △ be given, both in magnitude and position, and the sum of the squares of the sides in magnitude, the locus of the vertex is a ⊙.

Prop. 3.—*The sum of the squares of the diagonals of a parallelogram is equal to the sum of the squares of its four sides.*

Dem.—Let ABCD be the ▭. Draw CE ∥ to BD; produce AD to meet CE. Now, AD = BC (xxxiv.), and DE = BC; ∴ AD = DE; hence (2) $AC^2 + CE^2 = 2AD^2 + 2DC^2$; but $CE^2 = BD^2$; ∴ $AC^2 + BD^2 = 2AD^2 + 2DC^2$ = sum of squares of the four sides of the parallelogram.

Prop. 4.—*The sum of the squares of the four sides of a quadrilateral is equal to the sum of the squares of its diagonals plus four times the square of the line joining the middle points of the diagonals.*

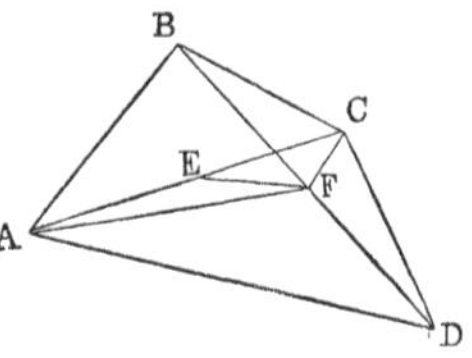

Dem.—Let ABCD be the quadrilateral, E, F the middle points of the diagonals. Now, in the △ ABD, $AB^2 + AD^2 = 2AF^2 + 2FB^2$, (2)

and in the △ BCD, $BC^2 + CD^2 = 2CF^2 + 2FB^2$; (2)

therefore $AB^2 + BC^2 + CD^2 + DA^2 = 2(AF^2 + CF^2) + 4FB^2$

$$= 4AE^2 + 4EF^2 + 4FB^2 = AC^2 + BD^2 + 4EF^2.$$

Prop. 5.—*Three times the sum of the squares of the sides of a triangle is equal to four times the sum of the squares of the lines bisecting the sides of the triangle.*

Dem.—Let D, E, F be the middle points of the sides.

Then $AB^2 + AC^2 = 2BD^2 + 2DA^2$; (2)
therefore $2AB^2 + 2AC^2 = 4BD^2 + 4DA^2$;
that is $2AB^2 + 2AC^2 = BC^2 + 4DA^2$.
Similarly $2BC^2 + 2BA^2 = CA^2 + 4EB^2$;
and $2CA^2 + 2CB^2 = AB^2 + 4FC^2$.
Hence $3(AB^2 + BC^2 + CA^2) = 4(AD^2 + BE^2 + CF^2)$.

Cor. If G be the point of intersection of the bisectors of the sides, $3AG = 2AD$; hence $9AG^2 = 4AD^2$;

$$\therefore\ 3(AB^2 + BC^2 + CA^2) = 9(AG^2 + BG^2 + CG^2);$$
$$\therefore\ (AB^3 + BC^2 + CA^2) = 3(AG^2 + BG^2 + CG^2).$$

Prop. 6.—*The rectangle contained by the sum and difference of two sides of a triangle is equal to twice the rectangle contained by the base, and the intercept between the middle point of the base and the foot of the perpendicular from the vertical angle on the base* (see Fig., Prop. 2).

Let CE be the ⊥ and D the middle point of the base AB.

Then $AC^2 = AE^2 + EC^2$,
and $BC^2 = BE^2 + EC^2$;
therefore, $AC^2 - BC^2 = AE^2 - EB^2$;
or $(AC + BC)(AC - BC) = (AE + EB)(AE - EB)$.
Now, $AE + EB = AB$, and $AE - EB = 2ED$;
therefore $(AC + BC)(AC - BC) = 2AB \cdot ED$.

Prop. 7.—*If* A, B, C, D *be four points taken in order on a right line, then* AB . CD + BC . AD = AC . BD.

A B C D

Dem.—Let $AB = a$, $BC = b$, $CD = c$; then $AB \cdot CD + BC \cdot AD = ac + b(a + b + c) = (a + b)(b + c) = AC \cdot BD$.

This theorem, which is due to Euler, is one of the most important in Elementary Geometry. It may be written in a more symmetrical form by making use of the convention regarding plus and minus : thus, since $+ AC = - CA$, we get

$$AB \,.\, CD + BC \,.\, AD = - CA \,.\, BD,$$

or

$$AB \,.\, CD + BC \,.\, AD + CA \,.\, BD = 0.$$

Prop. 8.—*If perpendiculars be drawn from the angular points of a square to any line, the sum of the squares of the perpendiculars from one pair of opposite angles exceeds twice the rectangle of the perpendiculars from the other pair of opposite angles by the area of the square.*

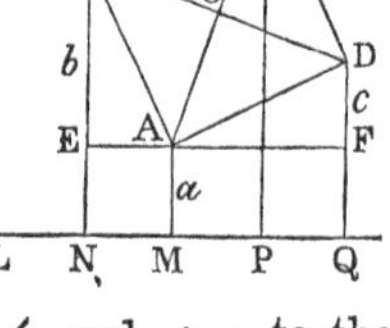

Dem. — Let ABCD be the square, L the line; let fall the ⊥s AM, BN, CP, DQ, on L: through A draw EF ∥ to L. Now, since the ∠ BAD is right, the sum of the ∠s BAE, DAF = one right ∠, and ∴ = to the sum of the ∠s BAE, ABE; ∴ ∠ ABE = DAF, and ∠ E = F, and AB = AD ; ∴ AE = DF.

Again, put AM = a, BE = b, DF = c. The four ⊥s can be expressed in terms of a, b, c. For BN = $a + b$, DQ = $a + c$; and since O is the middle point both of AC and BD, we have BN + DQ = AM + CP, each being = twice the ⊥ from O. Hence $(a + b) + (a + c) = a + CP$;

therefore $$CP = (a + b + c).$$

Now, $$BN^2 + DQ^2 - 2AM \,.\, CP = (a + b)^2 + (a + c)^2 - 2a(a + b + c) = b^2 + c^2 = BE^2 + DF^2.$$
$$= BE^2 + EA^2 = BA^2 = \text{area of square.}$$

Prop. 9.—*If the base* AB *of a triangle be divided in* D, *so that* m AD = n DB ; *then* $mAC^2 + nBC^2 = mAD^2 + nBD^2 + (m + n)\,CD^2$.

Dem.—Let fall the ⊥ CE; then

$$mAC^2 = m(AD^2 + DC^2 + 2AD \,.\, DE) ; \qquad \text{(xii.)}$$
$$nBC^2 = n\,(BD^2 + DC^2 - 2DB \,.\, DE). \qquad \text{(xiii.)}$$

Now, since $m\text{AD} = n\text{DB}$, we have

$$m(2\text{AD}\,.\,\text{DE}) = n(2\text{DB}\,.\,\text{DE}).$$

Hence, by addition, the rectangles disappear, and we get

$$m\text{AC}^2 + n\text{BC}^2 = m\text{AD}^2 + n\text{BD}^2 + (m+n)\,\text{CD}^2.$$

Cor.—If the point D be in the line AB produced, and if $m\text{AD} = n\text{BD}$, we shall have

$$m\text{AC}^2 - n\text{BC}^2 = m\text{AD}^2 - n\text{DB}^2 + (m-n)\,\text{CD}^2.$$

This case is included in the last, if we consider that DB changes sign when the point D passes through B.

Prop. 10.—*If* A, B, C, D, &c., *be any system of n points,* O *their centre of mean position,* P *any other point, the sum of the squares of the distances of the points* A, B, C, D, &c., *from* P *exceeds the sum of the squares of their distances from* O *by* $n\text{OP}^2$.

Dem.—For the sake of simplicity, let us take four points, A, B, C, D. The method of proof is perfectly general, and can be extended to any number of points. Let M be the middle point of AB; join MC, and divide it in NC, so that MN $= \frac{1}{2}$NC; join ND, and divide in O, so that NO $= \frac{1}{3}$OD; then O is the centre of mean position of the four points A, B, C, D.

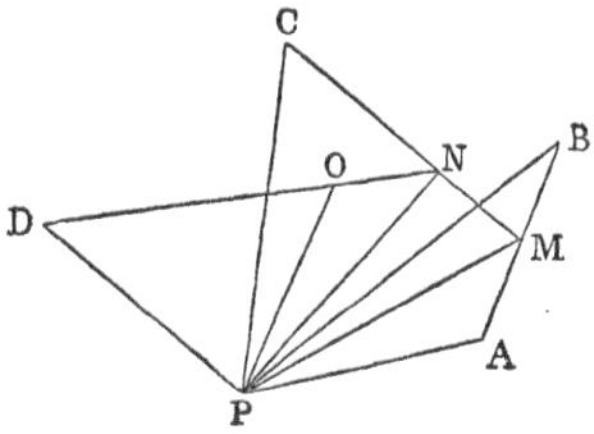

Now, applying the theorem of the last article to the several $\triangle$s APB, MPC, NPD, we have

$$\text{AP}^2 + \text{BP}^2 = \text{AM}^2 + \text{MB}^2 + 2\text{MP}^2;$$
$$2\text{MP}^2 + \text{CP}^2 = 2\text{MN}^2 + \text{NC}^2 + 3\text{NP}^2;$$
$$3\text{NP}^2 + \text{DP}^2 = 3\text{NO}^2 + \text{OD}^2 + 4\text{OP}^2.$$

Hence, by addition, and omitting terms that cancel on both sides, we get

$$\text{AP}^2 + \text{BP}^2 + \text{CP}^2 + \text{DP}^2 = \text{AM}^2 + \text{MB}^2 + 2\text{MN}^2 + \text{NC}^2 + 3\text{NO}^2 + \text{OD}^2 + 4\text{OP}^2.$$

Now, if the point P coincide with O, OP vanishes, and we have

$$AO^2 + BO^2 + CO^2 + DO^2 = AM^2 + MB^2 + 2MN^2 + NC^2 + 3NO^2 + OD^2;$$

therefore, $AP^2 + BP^2 + CP^2 + DP^2$
exceeds $AO^2 + BO^2 + CO^2 + DO^2$ by $4OP^2$.

Cor.—If O be the point of intersection of bisectors of the sides of a △, and P any other point; then

$$AP^2 + BP^2 + CP^2 = AO^2 + BO^2 + CO^2 + 3OP^2:$$

for the point of intersection of the bisectors of the sides is the centre of mean position.

Prop. 11.—*The last Proposition may be generalized thus: if* A, B, C, D, &c., *be any system of points,* O *their centre of mean position for any system of multiples* a, b, c, d, &c., then

$$a \,.\, AP^2 + b \,.\, BP^2 + c \,.\, CP^2 + d \,.\, DP^2, \text{ \&c.,}$$

exceeds $a \,.\, AO^2 + b \,.\, BO^2 + c \,.\, CO^2 + d \,.\, DO^2$, &c.,

by $(a + b + c + d, \text{ \&c.})\, OP^2$.

The foregoing proof may evidently be applied to this Proposition. The following is another proof from Townsend's *Modern Geometry*:—

From the points A, B, C, D, &c., let fall ⊥s AA′, BB′, CC′, DD′, &c., on the line OP; then it is easy to see that O is the centre of mean position for the points A′, B′, C′, D′, and the system of multiples a, b, c, d, &c.

Now we have by Props. xii., xiii., Book II.,

$$AP^2 = AO^2 + OP^2 + 2A'O \,.\, OP;$$
$$BP^2 = BO^2 + OP^2 + 2B'O \,.\, OP;$$
$$CP^2 = CO^2 + OP^2 + 2C'O \,.\, OP;$$
$$DP^2 = DO^2 + OP^2 + 2D'O \,.\, OP, \text{ \&c.};$$

therefore, multiplying by a, b, c, d, and adding, and remembering that

$$a \,.\, A'O + b \,.\, B'O + c \,.\, C'O + d \,.\, D'O + \text{\&c.} = 0 \text{ (see I., 18),}$$

we get

$$a \,.\, AP^2 + b \,.\, BP^2 + c \,.\, CP^2 + d \,.\, DP^2, \text{ \&c.,}$$
$$= a \,.\, AO^2 + b \,.\, BO^2 + c \,.\, CO^2 + d \,.\, DO^2 + \text{\&c.,}$$
$$+ (a + b + c + d, \text{ \&c.}) OP^2.$$

This Proposition evidently includes the last.

Cor. 1.—The locus of a point, the sum of the squares of whose distances from any number of given points, multiplied respectively by any system of constants a, b, c, d, is a circle, whose centre is the centre of mean position of the given points for the system of multiples a, b, c, d.

Cor. 2.—The sum of the squares for any system of multiples will be a minimum when the lines are drawn to the centre of mean position.

Prop. 12.—*From the* Propositions vi. and ix. *it follows that, if a line is divided into any two parts, the rectangle of the parts is a maximum, and the sum of their squares is a minimum, when the parts are equal.*

Cor.—If a line be divided into any number of parts, the continued product of all the parts is a maximum, and the sum of their squares is a minimum when they are all equal. For if we make any two of the parts unequal, we diminish the continued product, and we increase the sum of the squares.

SECTION II.

Exercises.

1. The second and third Propositions of the Second Book are special cases of the First.

2. Prove the fourth Proposition by the second and third.

3. Prove the sixth by the fifth, and the tenth by the ninth.

4. If the ∠ C of a △ ACB be $\frac{1}{3}$ of two right ∠ s, prove

$$AB^2 = AC^2 + CB^2 - AC \,.\, CD.$$

5. If C be $\frac{2}{3}$ of two right ∠ s, prove

$$AB^2 = AC^2 + CB^2 + AC \,.\, CB.$$

6. In a quadrilateral the sum of the squares of two opposite sides, together with the sum of the squares of the diagonals, is equal to the sum of the squares of the two remaining sides, together with four times the square of the line joining their middle points.

7. Divide a given line AB in C, so that the rectangle under BC and a given line may be equal to the square of AC.

8. Being given the rectangle contained by two lines, and the difference of their squares: construct them.

9. Produce a given line AB to C, so that AC . CB is equal to the square of another given line.

10. If a line AB be divided in C, so that $AB . BC = AC^2$, prove $AB^2 + BC^2 = 3AC^2$, and $(AB + BC)^2 = 5AC^2$.

11. In the fig. of Prop. xi. prove that—

(1). The lines GB, DF, AK, are parallel.

(2). The square of the diameter of the ⊙ about the △ FHK $= 6HK^2$.

(3). The square of the diameter of the ⊙ about the △ FHD $= 6FD^2$.

(4). The square of the diameter of the ⊙ about the △ AHD $= 6AH^2$.

(5). If the lines EB, CH intersect in J, AJ is ⊥ to CH.

12. If ABC be an isosceles △, and DE be ∥ to the base BC, and BE joined, $BE^2 - CE^2 = BC . DE$.

13. If squares be described on the three sides of any △, and the adjacent angular points of the squares joined, the sum of the squares of the three joining lines is equal to three times the sum of the squares of the sides of the triangle.

14. Given the base AB of a △, both in position and magnitude, and $mAC^2 - nBC^2$: find the locus of C.

15. If from a fixed point P two lines PA, PB, at right angles to each other, cut a given ⊙ in the points A, B, the locus of the middle point of AB is a ⊙.

16. If CD be any line ∥ to the diameter AB of a semicircle, and if P be any point in AB, then

$$CP^2 + PD^2 = AP^2 + PB^2.$$

17. If O be the mean centre of a system of points A, B, C, D, &c., for a system of multiples *a*, *b*, *c*, *d*, &c.; then, if L and M be any two ∥ lines,

$$\Sigma (a . AL^2) - \Sigma (a . AM^2) = \Sigma (a) . (OL^2 - OM^2).$$

BOOK THIRD.

SECTION I.

Additional Propositions.

Prop. 1.—*The two tangents drawn to a circle from any external point are equal.*

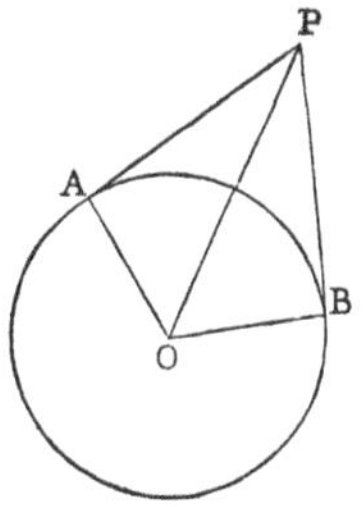

Dem.—Let PA, PB be the tangents, O the centre of the ⊙. Join OA, OP, OB; then

$$OP^2 = OA^2 + AP^2,$$
$$OP^2 = OB^2 + BP;$$

but $OA^2 = OB^2$; $\therefore AP^2 = BP^2$, and $AP = BP$.

Prop. 2.—*If two circles touch at a point* P, *and from* P *any two lines* PAB, PCD *be drawn, cutting the circles in the points* A, B, C, D, *the lines* AC, BD *joining the points of section are parallel.*

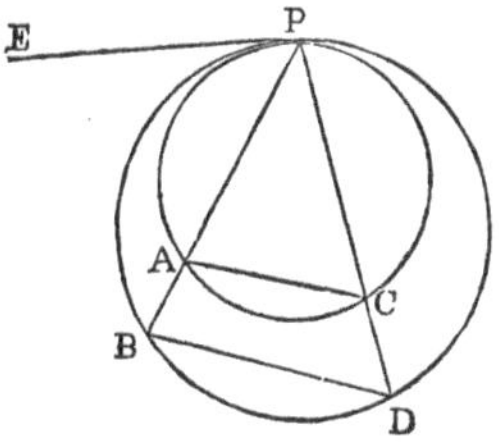

Dem.—At P draw the common tangent PE to both ⊙s; then

$\angle$ EPA = PCA; (xxxii.)
$\angle$ EPB = PDB.

Hence $\angle$ PCA = PDB, and AC is ∥ to BD. (I. xxiii.)

Cor.—If the angle APC be a right angle, AC and BD will be diameters of the ⊙s, and then we have the following *important* theorem. The lines drawn from the point of contact of two touching circles to the extremities of any diameter of one of them, will meet the other in points which will be the extremities of a parallel diameter.

Prop. 3.—*If two circles touch at* P, *and any line* PAB *cut both circles in* A *and* B, *the tangents at* A *and* B *are parallel.*

Dem.—Let the tangents at A and B meet the tangents at P in the points E and F.

Now, since AE = EP (1), the ∠ APE = PAE. In like manner, the ∠ BPF = PBF; ∴ ∠ PAE = PBF, and AE is ∥ to BF.

This Prop. may be inferred from (2), by supposing the lines PAB, PCD to approach each other indefinitely; then AC and BD will be tangents.

Prop. 4.—*If two circles touch each other at any point P, and any line cut the circles in the points* A, B, C, D; *then the angle* APB = CPD.

Dem.—Draw a tangent PE at P; then

∠ EPB = PCB; (xxxii.)
∠ EPA = PDA.

Hence, by subtraction, ∠ APB = CPD.

Prop. 5.—*If a circle touch a semicircle in* D *and its diameter in* P, *and* PE *be perpendicular to the diameter at* P, *the square on* PE *is equal to twice the rectangle contained by the radii of the circles.*

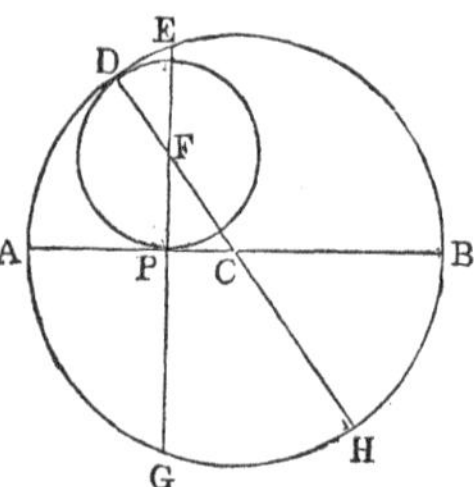

Dem.—Complete the circle, and produce EP to meet it again in G. Let C and F be the centres; then the line CF will pass through D. Let it meet the outside circle again in H.

Now, EF . FG = DF . FH (xxxv.), and $PF^2 = DF^2$. Hence, by addition, making use of II. v., and II. iii., EP^2 = DF . DH = twice rectangle contained by the radii.

Prop. 6.—*If a circle* PGD *touch a circle* ABC *in* D *and a chord* AB *in* P, *and if* EF *be perpendicular to* AB *at its middle point, and at the side opposite to that of the circle* PGD, *the rectangle contained by* EF *and the diameter of the circle* PGD *is equal to the rectangle* AP . PB.

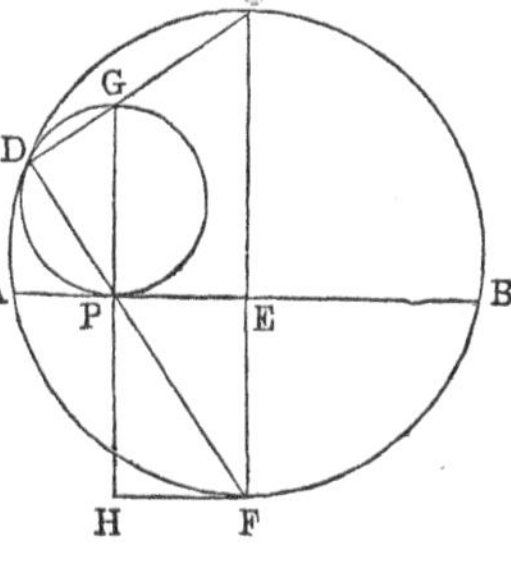

Dem.—Let PG be at right ∠s to AB, then PG is the diameter of the ⊙ PGD. Join DG, DP, and produce them to meet the ⊙ ABC in C and F; then CF is the diameter of the ⊙ ABC, and is ∥ to PG (2); ∴ CF is ⊥ to AB; hence it bisects AB in E (iii.). Through F draw FH ∥ to AB, and produce GP to meet it in H.

Now, since the ∠s H and D are right ∠s, a semicircle described on GF will pass through the points D and H.

Hence HP . PG = FP . PD = AP . PB; (xxxv.)
but HP = EF; ∴ EF . PG = AP . PB.

This Prop. and its Demonstration will hold true when the ⊙s are external to each other.

Cor. If AB be the diameter of the ⊙ ABC, this Prop. reduces to the last.

Prop. 7.—*To draw a common tangent to two circles.*

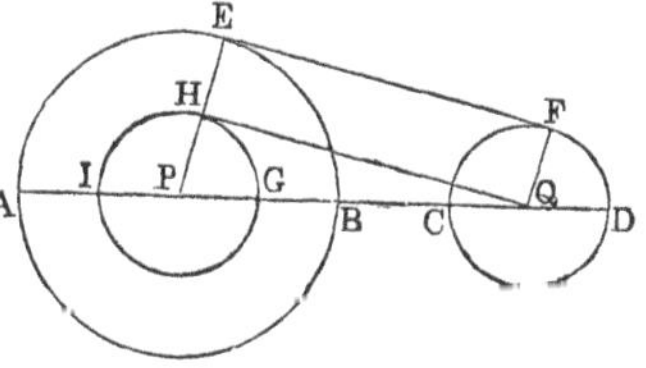

Let P be the centre of the greater ⊙, Q the centre of the less, with P as centre, and a radius = to the difference of the radii of the two ⊙s: describe the ⊙ IGH; from Q draw a tangent to this ⊙, touching it at H.

Join PH, and produce it to meet the circumference of the larger ⊙ in E. Draw QF ∥ to PE. Join EF, which will be the common tangent required.

Dem.—The lines HE and QF are, from the construction, equal; and since they are ∥, the fig. HEFQ is a ▭; ∴ the ∠ PEF = PHQ = right angle; ∴ EF is a tangent at E; and since ∠ EFQ = EHQ = right angle, EF is a tangent at F. The tangent EF is called a *direct* common tangent.

If with P as centre, and a radius equal to the sum of the radii of the two given ⊙s, we shall describe a ⊙, we shall have a common tangent which will pass between the ⊙s, and one which is called a transverse common tangent.

Prop. 8.—*If a line passing through the centres of two circles cut them in the points* A, B, C, D, *respectively; then the square of their direct common tangent is equal to the rectangle* AC . BD.

Dem.—We have (see last fig.) AI = CQ; to each add IC, and we get AC = IQ. In like manner, BD = GQ. Hence AC . BD = IQ . QG = EF^2.

Cor. 1.—If the two ⊙s touch, the square of their common tangent is equal to the rectangle contained by their diameters.

Cor. 2.—The square of the transverse common tangent of the two ⊙s = AD . BC.

Cor. 3.—If ABC be a semicircle, PE a ⊥ to AB from any point P, CQD a ⊙ touching PE, the semicircle ACB, and the semicircle on PB; then, if QR be the diameter of CQD, AB . QR = EP^2.

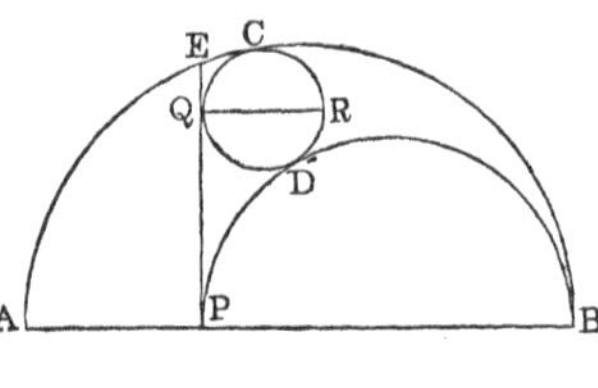

Dem. PB . QR = PQ^2, (*Cor.* 1)

AP . QR = $EP^2 - PQ^2$; (6)

therefore, by addition, AB . QR = EP^2.

Cor. 4.—If two ⊙s be described to touch an ordinate of a semicircle, the semicircle itself and the semicircles on the segments of the diameter, they will be equal to one another.

Prop. 9.—*In equiangular triangles the rectangles under the non-corresponding sides about equal angles are equal to one another.*

Dem.—Let the equiangular △s be ABO, DCO, and let them be placed so that the equal ∠s at O may be vertically opposite, and that the non-corresponding sides AO, CO may be in one right line, then the non-corresponding sides BO, OD shall be in one right line. Now, since the ∠ ABD = ACD, the four points A, B, C, D are concyclic (in the circumference of the same ⊙). Hence the rectangle AO . OC = rectangle BO . OD. (xxxv.)

Prop. 10.—*The rectangle contained by the perpendiculars from any point* O *in the circumference of a circle on two tangents* AC, BC, *is equal to the square of the perpendicular from the same point on their chord of contact* AB.

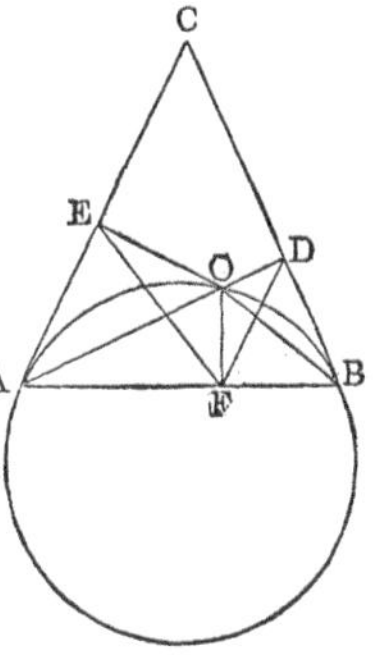

Dem.—Let the ⊥s be OD, OE, OF. Join OA, OB, EF, DF. Now, since the ∠s ODB, OFB, are right, the quadrilateral ODBF is inscribed in a ⊙. In like manner, the quadrilateral OEAF is inscribed in a ⊙. Again, since BC is a tangent, the ∠ DBO = BAO (xxxii.); but DBO = DFO (xxi.); and FAO = FEO; ∴ ∠ DFO = FEO. In like manner, ∠ ODF = EFO; hence the △s ODF, FEO are equiangular, and ∴ the rectangles contained by the non-corresponding sides about the equal ∠s DOF, FOE, are equal (9). Hence OD . OE = OF^2.

Prop. 11.—*If from any point* O *in the circumference of a circle perpendiculars be drawn to the four sides, and to* *the diagonals of an inscribed quadrilateral, the rectangle contained by the perpendiculars on either pair of opposite sides is equal to the rectangle contained by the perpendiculars on the diagonals.*

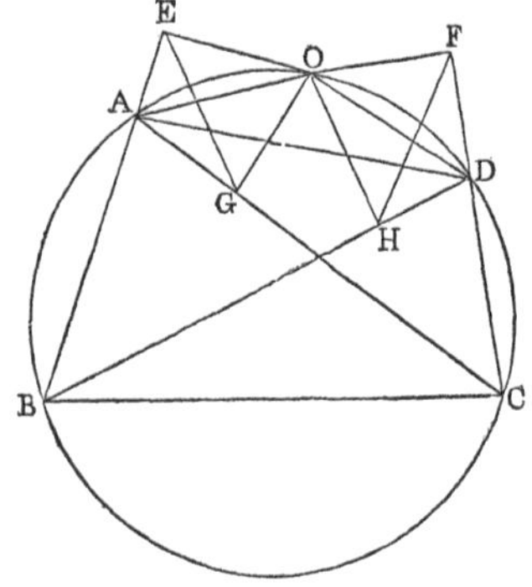

Dem.—Let OE, OF be the ⊥s on the opposite sides AB, CD; OG, OH, the ⊥s on the diagonals. Join EG, FH, OA, OD. Now, as in the last Prop., we see that the quadrilaterals AEOG, DFOH, are inscribed in ⊙s. Hence ∠ OEG = OAG, and OHF = ODF. Again, since AODC is a quadrilateral in a ⊙, the ∠ OAC + ODC = two right ∠s (xxii.) = ODC + ODF; ∴ the ∠ OAC = ODF. Hence the ∠ OEG = OHF. In like manner, the ∠ OGE = OFH. Hence the △s OEG, OHF are equiangular, and the rectangle OE . OF = the rectangle OG . OH.

Cor. 1.—The rectangle contained by the ⊥s on one pair of opposite sides is equal to the rectangle contained by the ⊥s on the other pair of opposite sides. This may be proved directly, or it follows at once from the theorem in the text.

Cor. 2.—If we suppose the points A, B, to become consecutive, and also the points C, D, then AB, CD become tangents; and from the theorem of this Article we may infer the theorem of Prop. 10.

Prop 12.—*The feet* D, E, F *of the three perpendiculars let fall on the sides of a triangle* ABC, *from any point* P *in the circumference of the circumscribed circle, are collinear.*

Dem.—Join PA, PB, DF, EF. As in the Demonstrations of the two last Propositions, we see that the quadrilaterals PBDF, PFAE are inscribed in ⊙s; ∴ the ∠s PBD, PFD are = two right ∠s (xxii.), and ∠s PBD,

PAC, are = two right ∠s (xxii.); ∴ ∠ PFD = PAC; and since PFAE is a quadrilateral in a circle, the ∠ EAP = EFP; ∴ PFD + PFE = PAC + PAE = two right ∠s. Hence the points D, F, E, are collinear.

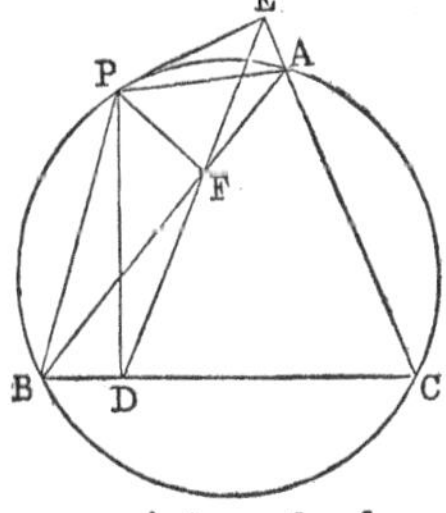

Cor. 1.—If the feet of the ⊥s drawn from any point P to the sides of the △ ABC be collinear, the locus of P is the ⊙ described about the triangle.

Cor. 2.—If four lines be given, a point can be found such, that the feet of the four ⊥s from it on the lines will be collinear. For let the four lines be AB, AC, DB, DF. These lines form four △s. Let the ⊙s described about two of the △s—say AFE, CDE—intersect in P; then it is evident that the feet of the ⊥s from P on the four lines will be collinear.

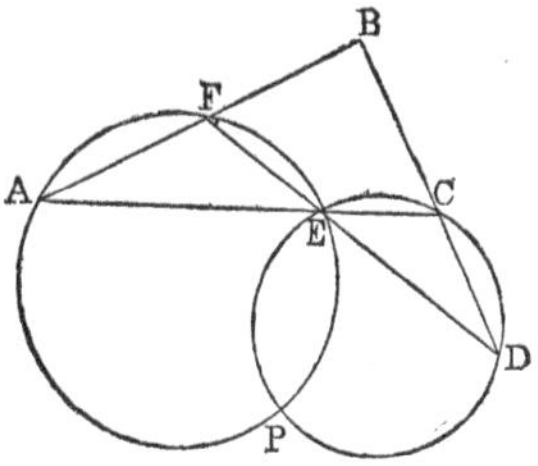

Cor. 3.—The ⊙s described about the △s ABC, DBF, each passes through the point P. This follows because the feet of the ⊥s from P on the sides of these △s are collinear.

Prop. 13.—*If the perpendiculars of a triangle be produced to meet the circumference of the circumscribed circle, the parts of the perpendiculars intercepted between their point of intersection and the circumference are bisected by the sides of the triangle.*

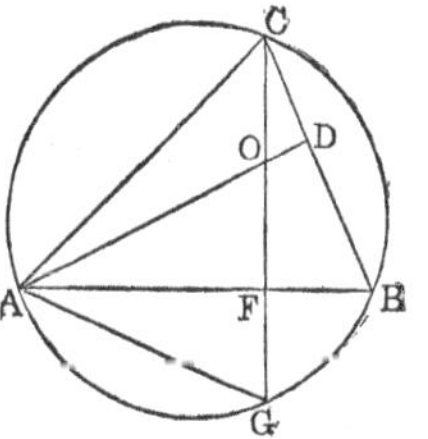

Let AD, CF intersect in O; produce CF to meet the ⊙ in G; then OF = FG.

Dem.—The ∠ AOF = COD (I. xv.) and AFO = CDO,

each being right; ∴ FAO = OCD; but OCD = GAF (xxi.); ∴ FAO = FAG, and AFO = AFG, each being right, and AF common. Hence OF = FG.

Prop. 14.—*The line joining any point* P, *in the circumference of a circle, to the point of intersection of the perpendiculars of an inscribed triangle, is bisected by the line of collinearity of the feet of the perpendiculars from* P *on the sides of the triangle.*

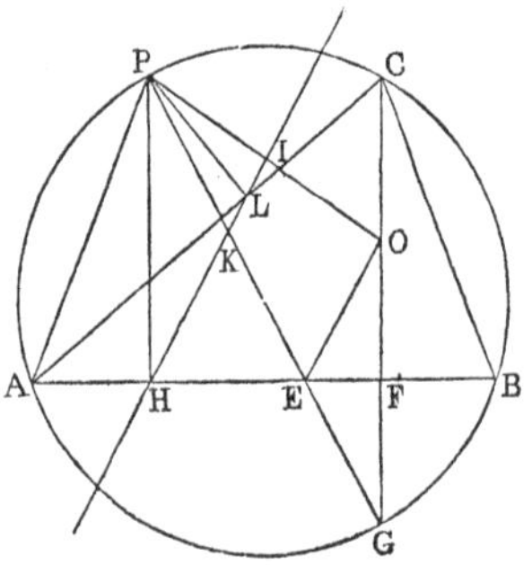

Let P be the point; PH, PL two of the ⊥s from P on the sides; thus HL is the line of collinearity of the feet of the ⊥s from P on the sides of the △. Let CF be the ⊥ from C on AB; produce CF to G, and make OF = FG; then O is the point of intersection of the ⊥s of the △. Join OP, intersecting HL in I: it is required to prove that OP is bisected in I.

Dem.—Join AP, PG, and let PG intersect HL in K, and AB in E. Join OE. Now, since APLH is a quadrilateral in a ⊙, the ∠ PHK = PAC = PGC = HPK; ∴ PK = KH. Hence KH = KE, and PK = KE. Again, since OF = FG, and FE common, ∠ GEF = OEF; but GEF = KEH = KHE; ∴ ∠ OEF = KHE; ∴ OE is ∥ to KH; and since EP is bisected in K, OP is bisected in I.

Cor.—If X, Y, Z, W be the points of intersection of the ⊥s of the four △s AFE, CDE, ABC, DBF (see fig., *Cor.* 2, Prop. 12), then X, Y, Z, W are collinear. For let L denote the line of collinearity of the feet of the ⊥s from P on the sides of the four △s. Join PX, PY, PZ, PW. Then, since L joins the points of bisection of the sides of the △ PXY, the line XY is ∥ to L. Similarly, YZ, ZW are each ∥ to L. Hence XY, YZ, ZW form one continuous line.

Prop. 15.—*Through one of the points of intersection of two given circles to draw a line, the sum of whose segments intercepted by the circles shall be a maximum.*

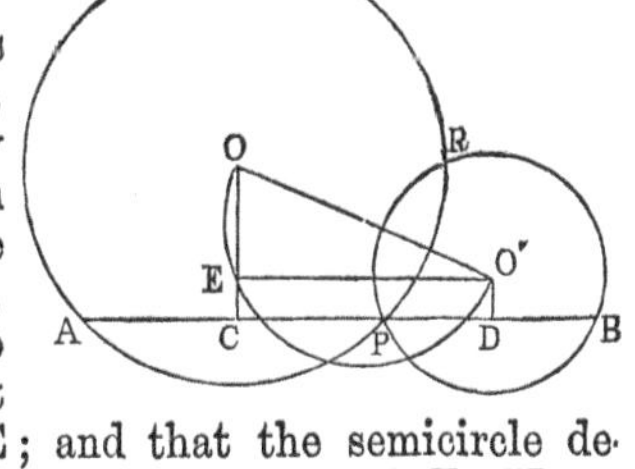

Analysis.—Let the ⊙s intersect in the points P, R, and let APB be any line through P. From O, O′, the centres of the ⊙s, let fall the ⊥s OC, O′D, and draw O′E ∥ to AB. Now, it is evident that AB = 2CD = 2O′E; and that the semicircle described on OO′ as diameter will pass through E. Hence it follows that if AB is a maximum, the chord O′E will coincide with OO′. Therefore AB must be ∥ to the line joining the centres of the ⊙s.

Cor. 1.—If it were required to draw through P a line such that the sum of the segments AP, PB may be equal to a given line, we have only to describe a ⊙ from O′ as centre, with a line equal half the given line as radius; and the place where this ⊙ intersects the ⊙ on OO′ as diameter will determine the point E; and then through P draw a ∥ to O′E.

Def.—*A triangle is said to be given in species when its angles are given.*

Prop. 16.—*To describe a triangle of given species whose sides shall pass through three given points, and whose area shall be a maximum.*

Analysis.—Let A, B, C be the given points, DEF the required △; then, since the triangle DEF is given in species, the ∠s D, E, F are given, and the lines AB, BC, CA are given by hypothesis; ∴ the ⊙s about the △s ABF, BCD, CAE are given. These three ⊙s will intersect in a common point. For, let the two first intersect in O. Join AO, BO, CO; then ∠AFB + AOB = two right ∠s; and BDC + BOC = two right ∠s; ∴ the ∠s AFB, BDC, AOB, COB = four right ∠s, and the ∠s

AOB, BOC, COA = four right ∠s; ∴ the ∠ COA = AFB + BDC: to each add the ∠ CEA, and we have the ∠ COA + CEA = sum of the three ∠s of the △ DEF, that is = two right ∠s; ∴ the quadrilateral AECO is inscribed in a ⊙. Hence the three ⊙s pass through a common point, which is a given point.

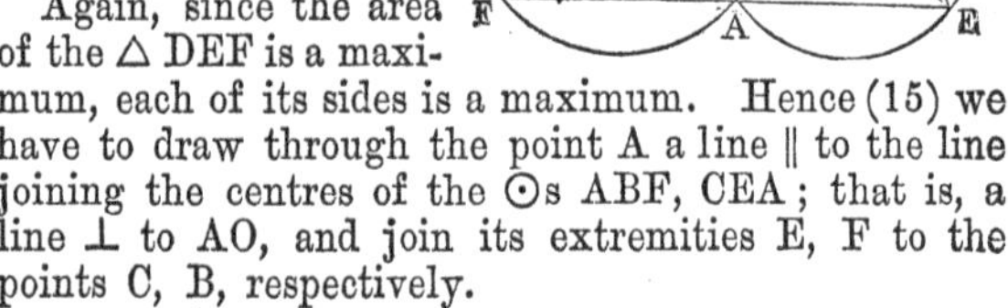

Again, since the area of the △ DEF is a maximum, each of its sides is a maximum. Hence (15) we have to draw through the point A a line ∥ to the line joining the centres of the ⊙s ABF, CEA; that is, a line ⊥ to AO, and join its extremities E, F to the points C, B, respectively.

Cor.—If instead of the maximum △ we require to describe a △ whose sides will be equal to three given lines, the method of solving the question can be inferred from the corollary to the last Proposition.

Prop. 17.—*To describe in a given triangle* DEF (see last fig.) *a triangle given in species whose area shall be a minimum.*

Analysis.—Let ABC be the inscribed △; describe ⊙s about the three △s ABF, BCD, CAE; then these ⊙s will have a common point: let it be O. We prove this to be a given point as follows: The ∠ FOE exceeds the ∠ FDE by the sum of the ∠s DFO, DEO; that is, by the sum of the ∠s BAO, CAO. Hence the ∠ FOE = FDE + BAC; ∴ the ∠ FOE is given. In like manner, the ∠ EOD is given. Hence the point O will be the point of intersection of two given ⊙s, and is ∴ given; and, since E and F are given points, the ∠ OFE is given; ∴ the ∠ OBA is given. In like manner, the ∠ OAB is given; ∴ △ OAB is given in species. Now, since the △ ABC is a minimum, the side AB is a minimum; ∴ OA is a minimum; and since O is a given point, OA must be ⊥ to EF. Hence the method of inscribing the minimum △ has been found.

Cor.—From the foregoing analysis the method is obvious of inscribing in a given △ another △ whose sides shall be respectively equal to three given right lines.

Prop. 18.—*If* ABC *be a triangle, and* CD *a perpendicular to* AB; *then if* AE = DB, *it is required to prove that* AB *is the minimum line that can be drawn through* E, *meeting the two fixed lines* AC, BC.

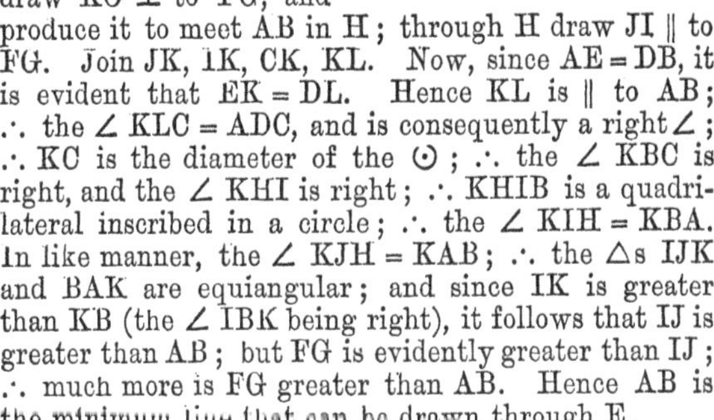

Dem.—Describe a ⊙ about the △ ABC; produce CD to meet it in L, and erect EK ⊥ to AB. Join AK, BK. Through E draw any other line FG; draw KO ⊥ to FG, and produce it to meet AB in H; through H draw JI ∥ to FG. Join JK, IK, CK, KL. Now, since AE = DB, it is evident that EK = DL. Hence KL is ∥ to AB; ∴ the ∠ KLC = ADC, and is consequently a right ∠; ∴ KC is the diameter of the ⊙; ∴ the ∠ KBC is right, and the ∠ KHI is right; ∴ KHIB is a quadrilateral inscribed in a circle; ∴ the ∠ KIH = KBA. In like manner, the ∠ KJH = KAB; ∴ the △s IJK and BAK are equiangular; and since IK is greater than KB (the ∠ IBK being right), it follows that IJ is greater than AB; but FG is evidently greater than IJ; ∴ much more is FG greater than AB. Hence AB is the minimum line that can be drawn through E.

If in the foregoing fig. the line BA receive an infinitely small change of position, namely, B along BC, and A along AC; then it is plain the motions of B and A would be the same as if the △ AKB got an infinitely small turn round the point K, which remains fixed: on this account the point K is called the centre of instantaneous rotation for the line AB.

This Proposition admits of another demonstration, as follows:—Through the points A, B draw the lines AM, BM ∥ to BC, AC; then ME is evidently ⊥ to AB; let fall the ⊥ MN on FG; join AG, MG; then the △ FMG is plainly greater than △ AGM; but △ AGM = △ ABM; ∴ △ FGM is greater than △ ABM, and its ⊥ MN is less than ME, the ⊥ of △ AMB; hence the base FG is greater than the base AB.

Prop. 19.—*If* OC, OD *be any two lines*, AB *any arc of a circle, or of any other curve concave to* O; *then, of all the tangents which can be drawn to* AB, *that whose intercept is bisected at the point of contact cuts off the minimum triangle.*

Dem.—Let CD be bisected at P, and let EF be any other tangent. Then through P draw GH ∥ to EF; then, since CD is bisected in P, the △ cut off by CD is less than the △ cut off by GH (I. 19); but the △ cut off by GH is less than the △ cut off by EF. Hence the △ cut off by CD is less than the △ cut off by EF.

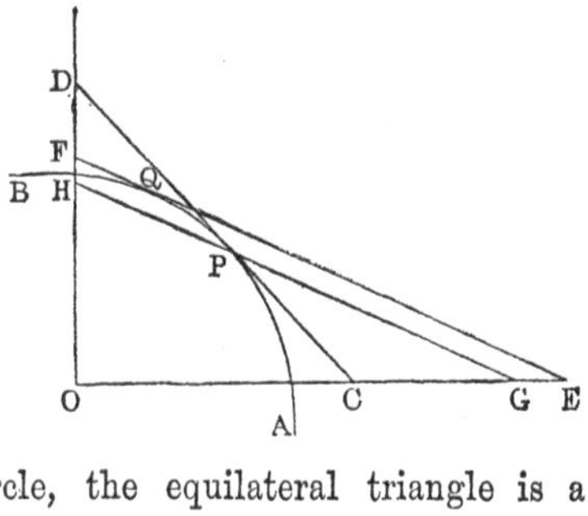

Cor. 1.—Of all triangles described about a given circle, the equilateral triangle is a minimum.

Cor. 2.—Of all polygons having a given number of sides described about a given ⊙, the regular polygon is a minimum.

Prop. 20.—*If* ABC *be a circle,* AB *a diameter,* PD *a fixed line perpendicular to* AB; *then if* ACP *be any line cutting the circle in* C *and the line* PD *in* P, *the rectangle under* AP *and* AC *is constant.*

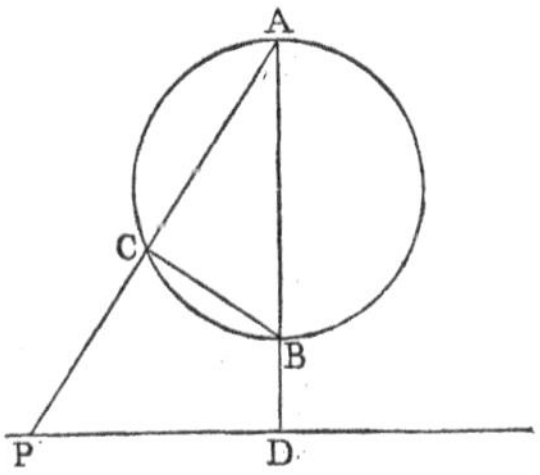

Dem.—Since AB is the diameter of the ⊙, the ∠ ACB is right (xxxi.); ∴ BCP is right, and BDP is right; ∴ the figure BDPC is a quadrilateral inscribed in a ⊙, and, consequently, the rectangle AP . AC = rectangle AB . AD = constant.

Cor. 1.—This Prop. holds true when the line PD cuts the ⊙, as in the diagram: the value of the constant will, in this case, be = AE^2. Hence we have the following:—

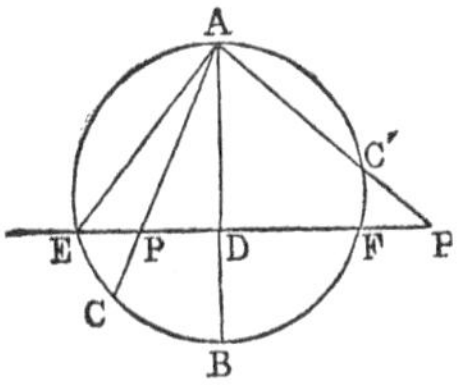

Cor. 2.—If A be the middle point of the arc EF, AC any chord cutting the line EF in P; then AP . AC = AE^2.

On account of its importance, we shall give an independent proof of this Prop. Thus: join EC, and suppose a ⊙ described about the △ EPC; then the ∠ FEA = ECA, because they stand on equal arcs AF, AE. Hence AE touches the ⊙ EPC (xxxii.); ∴ the rectangle AP . AC = AE^2.

Cor. 3.—If A be a fixed point (see two last figs.), PD a fixed line, and if any variable point P in PD be joined to A, and a point C taken on AP, so that the rectangle AP . AC = constant—say R^2—then, by the converse of this Prop., the locus of the point C is a ⊙.

Def.—*The point* C *is called the* inverse *of the point* P, *the* ⊙ ABC *the* inverse *of the line* PD, *the fixed point* A *the* centre, *and the constant* R *the* radius *of* inversion.

We shall give more on the subject of inversion in our addition to Book VI.

Prop. 21.—*If from the centre of a circle a perpendicular be let fall on any line* GD, *and from* D, *the foot of the perpendicular, and from any other point* G *in* GD *two tangents* DE, GF *be drawn to the circle, then* $GF^2 = GD^2 + DE^2$.

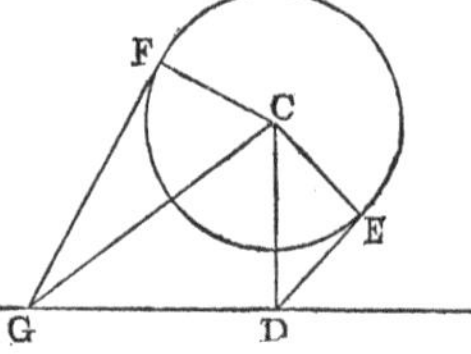

Dem.—Let C be the centre of the ⊙. Join CG, CE, CF. Then

$$GF^2 = GC^2 - CF^2 = GD^2 + DC^2 - CF^2$$
$$= GD^2 + DE^2 + EC^2 - CF^2 = GD^2 + DE^2.$$

Prop. 22.—*To describe a circle having its centre at a given point, and cutting a given circle orthogonally (at right angles).*

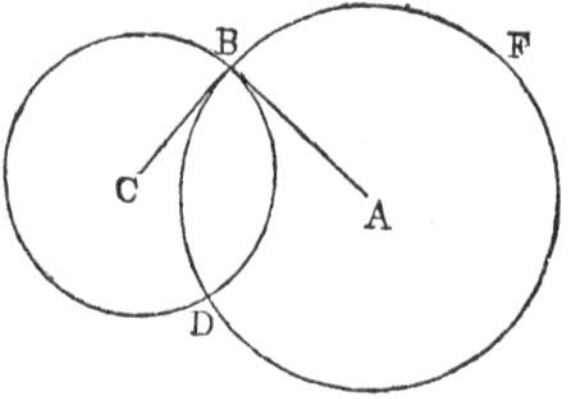

Let A be the given point, BED the given ⊙. From A draw AB, touching the ⊙ BED (xvii.) at B; and from A as centre, and AB as radius, describe the ⊙ BFD: this ⊙ will cut BED orthogonally.

Dem.—Let C be the centre of BED. Join CB; then, because AB is a tangent to the circle BED, CB is at right ∠s to AB (xviii.); ∴ CB touches the ⊙ BDF. Now, since AB, CB are tangents to the ⊙s BDE, BDF, these lines coincide with the ⊙s for an indefinitely short distance (a tangent to a ⊙ has two consecutive points common with the ⊙); and, since the lines intersect at right ∠s, the ⊙s cut at right ∠s; that is, orthogonally.

Cor. 1.—The ⊙s cut also orthogonally at D.

Cor. 2.—When two ⊙s cut orthogonally, the square of the distance between their centres is equal to the sum of the squares of their radii.

Prop. 23.—*If in the line joining the centres of two circles a point* D *be found, such that the tangents* DE, DE′ *from it to the circles are equal, and if through* D *a line* DG *be drawn perpendicular to the line joining the centres, then the tangents from any other point* G *in* DG *to the circles will be equal.*

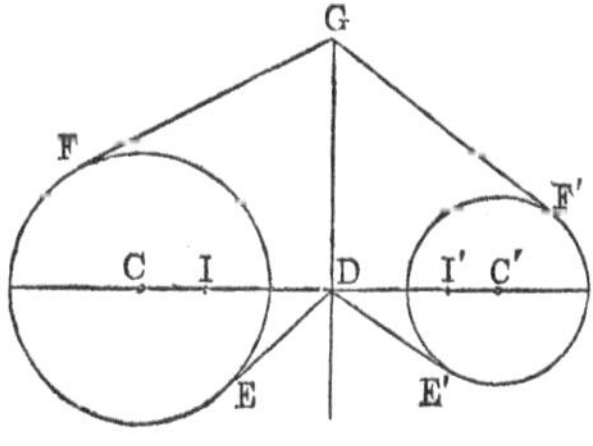

Dem.—Let GF, GF′ be the tangents. Now, by hypothesis, $DE^2 = DE'^2$. To each add DG^2, and we have

$$GD^2 + DE^2 = GD^2 + DE'^2,$$

or $$GF^2 = GF'^2; \therefore GF = GF'.$$

Def.—*The line* GD *is called the radical axis of the two circles; and two points* I, I′, *taken on the line through the centres, so that* DI = DI′ = DE = DE′, *are called the limiting points.*

Cor. 1.—Any circle whose centre is on the radical axis, and which cuts one of the given ⊙s orthogonally, will also cut the other orthogonally, and will pass through the two limiting points.

Cor. 2.—If there be a system of three ⊙s, their radical axes taken in pairs are concurrent. For, if tangents be drawn to the ⊙s from the point of intersection of two of the radical axes, the three tangents will be equal. Hence the third radical axis passes through this point.

Def.—*The point of concurrence of the three radical axes is called the radical centre of the circles.*

Cor. 3.—The ⊙ whose centre is the radical centre of three given ⊙s, and which cuts one of them orthogonally, cuts the other two orthogonally.

Prop. 24.—*The difference between the squares of the tangents, from any point* P *to two circles, is equal to twice the rectangle contained by the perpendicular from* P *on the radical axis and the distance between the centres of the circles.*

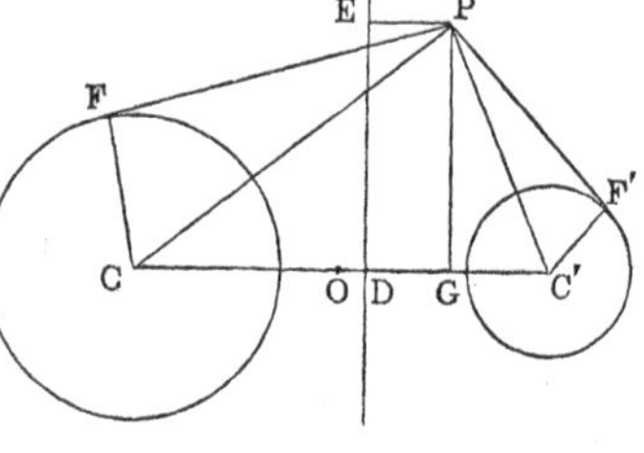

Dem.—Let C, C′, be the centres, O the middle point of CC′, DE the radical axis. Let fall the ⊥s PE, PG. Now,

$$CP^2 - C'P^2 = 2CC' . OG \qquad \text{(II., 6)}$$

$$CF^2 - C'F'^2 = CD^2 - C'D^2,$$

because DE is the radical axis

$$= 2CC' . OD.$$

Hence, by subtraction,

$$PF^2 - PF'^2 = 2CC' . DG = 2CC' . EP.$$

This is the fundamental Prop. in the theory of coaxal circles. For more on this subject, see Book VI., Section v.

Def.—*If on any radius of a circle two points be taken, one internally and the other externally, so that the rectangle contained by their distances from the centre is equal to the square of the radius; then a line drawn perpendicular to the radius through either point is called the* polar *of the other point, which is called, in relation to this perpendicular, its* pole. *Thus, let* O *be the centre, and let* OA . OP = *radius*^{2}; *then, if* AX, PY *be perpendiculars to the line* OP, PY *is called the* polar *of* A, *and* A *the* pole *of* PY. *Similarly,* AX *is the polar of* P, *and* P *the pole of* AX.

Prop. 25.—*If* A *and* B *be two points, such that the polar of* A *passes through* B, *then the polar of* B *passes through* A.

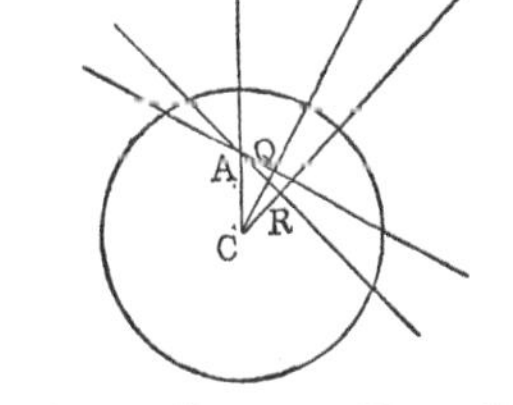

Dem.—Let the polar of A be the line PB; then PB is ⊥ to CP (C being the centre). Join CB, and let fall the ⊥ AQ on CB. Then, since the ∠s P and Q are right ∠s, the quadrilateral APBQ is inscribed in a ⊙; ∴ CQ . CB = CA . CP = radius^2; ∴ AQ is the polar of B.

Cor.—In PB take any other point D. Join CD, and let fall the perpendicular AR on CD. Then AQ, AR are the polars of the points B and D, and we see that the line BD, which joins the points B and D, is the polar of the point A; the intersection of AQ, AR, the polars of B and D. Hence we have the following important theorem:—*The line of connexion of any two points is the polar of the point of intersection of their polars;* or, again: *The point of intersection of any two lines is the pole of the line of connexion of their poles.*

Def.—*Two points, such as* A *and* B, *which possess the property that the polar of either passes through the other, are called* conjugate *points with respect to the circle, and their polars are called* conjugate *lines.*

Prop. 26.—*If two circles cut orthogonally, the extremities of any diameter of either are conjugate points with respect to the other.*

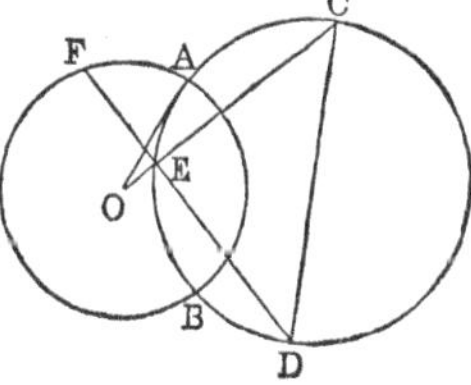

Let the ⊙s be ABF and CED, cutting orthogonally in the points A, B; let CD be any diameter of the ⊙ CED; C and D are conjugate points with respect to the ⊙ ABF.

Dem.—Let O be the centre of the ⊙ ABF. Join OC, intersecting the ⊙ CED in E. Join ED, and produce to F. Join OA. Now, because the ⊙s intersect orthogonally, OA is a tangent to the ⊙ CED. Hence OC . OE = OA^2; that is, OC . OE = square of radius of the ⊙ ABF; and, since the ∠ CED is a right angle, being in a semicircle, the line ED is the polar of C. Hence C and D are conjugate points with respect to the ⊙ ABF.

Prop. 27.—*If* A *and* B *be two points, and if from* A *we draw a perpendicular* AP *to the polar of* B, *and from* B *a perpendicular* BQ *to the polar of* A; *then, if* C *be the centre of the circle, the rectangle* CA . BQ = CB . AP (Salmon).

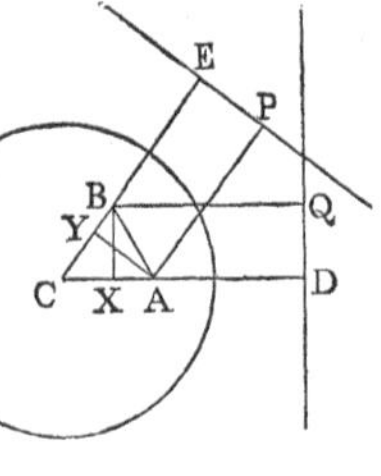

Dem.—Let fall the ⊥s AY, BX, on the lines CE, CD. Now, since X and Y are right angles, the semicircle on AB passes through the points X, Y.

Therefore CA . CX = CB . CY;

and CA . CD = CB . CE,

because each = radius^2; ∴ we get, by subtraction,

CA . DX = CB . EY;

or CA . BQ = CB . AP.

Prop. 28.—*The locus of the intersection of tangents to a circle, at the extremities of a chord which passes through a given point, is the polar of the point.*

Dem.—Let CD be the chord, A the given point, CE, DE the tangents. Join OA, and let fall the ⊥ EB on OA produced. Join OC, OD. Now, since EC = ED, and EO common, and OC = OD, the ∠ CEO = DEO. Again, since CE = DE, and EF common, and ∠ CEF

= DEF; ∴ the ∠ EFC = EFD. Hence each is right. Now, since the △ OCE is right-angled at C, and CF perpendicular to OE, OF . OE = OC^2; but since the quadrilateral AFEB has the opposite angles B and F right angles, it is inscribed in a ⊙. The rectangle OF . OE = OA . OB; but OF . OE = OC^2; ∴ OA . OB = OC^2 = radius^2; ∴ BE is the polar of A, and this is the locus of the point E.

Cor. 1.—If from every point in a given line tangents be drawn to a given circle, the chord of contact passes through the pole of the given line.

Cor. 2.—If from any given point two tangents be drawn to a given circle, the chord of contact is the polar of the given point.

Prop. 29.—The older geometers devoted much time to the solution of problems which required the construction of triangles under certain conditions. Three independent data are required for each problem. We give here a few specimens of the modes of investigation employed in such questions, and we shall give some additional ones under the Sixth Book.

(1). *Given the base of a triangle the vertical angle, and the sum of the sides: construct it.*

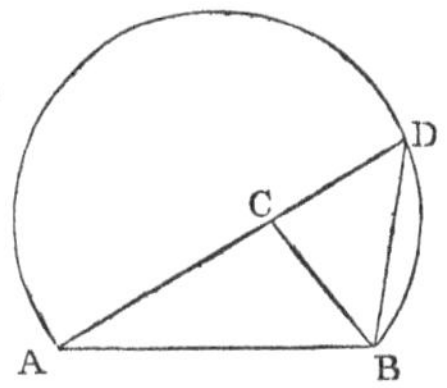

Analysis.—Let ABC be the △; produce AC to D, and make CD = CB; then AD = sum of sides, and is given; and the ∠ ADB = half the ∠ ACB, and is given. *Hence we have the following method of construction:*—On the base AB describe a segment of a ⊙ containing an ∠ = half the given vertical ∠, and from the centre A, with a distance equal to the sum of the sides as radius, describe a ⊙ cutting this segment in D. Join AD, DB, and make the ∠ DBC = ADB; then ABC is the △ required.

(2). *Given the vertical angle of a triangle, and the segments into which the line bisecting it divides the base: construct it.*

Analysis.—Let ABC be the △, CD the line bisecting the vertical ∠. Then AD, DB, and the ∠ ACB are given. Now, since AD, DB are given, AB is given; and since AB and the ∠ACB are given, the ⊙ ACB is given (xxxiii.); and since CD bisects the ∠ ACB, we have arc AE = EB; ∴ E is a given point, and D is a given point. Hence the line ED is given in position, and therefore the point C is given.

(3). *Given the base, the vertical angle, and the rectangle of the sides, construct the triangle.*

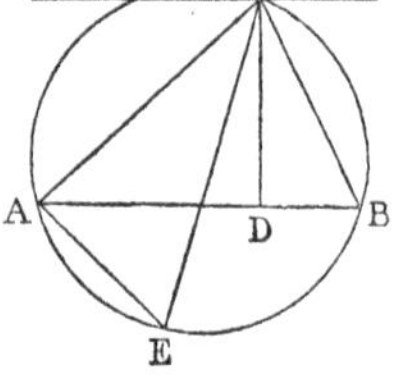

Analysis.—Let ABC be the △; let fall the ⊥ CD; draw the diameter CE; join AE. Now the ∠ CEA = CBA (xxi.), and CAE is right, being in a semicircle (xxxi.); ∴ = ∠ CDB. Hence the △s CAE, CDB are equiangular; ∴ rectangle AC . CB = rectangle CE . CD (9); but rectangle AC . CB is given; ∴ rectangle CE . CD is given; and since the base and vertical ∠ are given, the ⊙ ACB is given; ∴ the diameter CE is given; ∴ CD is given; and therefore the line drawn through C ∥ to AB is given in position. Hence the point C is given.

The method of construction is obvious.

SECTION II.

Exercises.

1. The line joining the centres of two ⊙s bisects their common chord perpendicularly.

2. If AB, CD be two ∥ chords in a ⊙, the arc AC = BD.

3. If two ⊙s be concentric, all tangents to the inner ⊙ which are terminated by the outer ⊙ are equal to one another.

4. If two ⊥s AD, BE of a △ intersect in O, AO . OD = BO . OE.

5. If O be the intersection of the ⊥s of a △, the ⊙s described about the three △s AOB, BOC, COA are equal to one another.

6. If equilateral △s be described on the three sides of any △, the ⊙s described about these equilateral △s pass through a common point.

7. The lines joining the vertices of the original △ to the opposite vertices of the equilateral △s are concurrent.

8. The centres of the three ⊙s in question 6 are the angular points of another equilateral △. This theorem will hold true if the equilateral △s on the sides of the original △ be turned inwards.

9. The sum of the squares of the sides of the two new equilateral △s in the last question is equal to the sum of the squares of the sides of the original triangle.

10. Find the locus of the points of bisection of a system of chords which pass through a fixed point.

11. If two chords of a ⊙ intersect at right angles, the sum of the squares of their four segments equal the square of the diameter.

12. If from any fixed point C a line CD be drawn to any point D in the circumference of a given ⊙, and a line DE be drawn ⊥ to CD, meeting the ⊙ again in E, the line EF drawn through E ∥ to CD will pass through a fixed point.

13. Given the base of a △ and the vertical ∠, prove that the sum of the squares of the sides is a maximum or a minimum when the △ is isosceles, according as the vertical ∠ is acute or obtuse.

14. Describe the maximum rectangle in a given segment of a circle.

15. Through a given point inside a ⊙ draw a chord which shall be divided as in Euclid, Prop. XI., Book II.

16. Given the base of a △ and the vertical ∠, what is the locus—(1) of the intersection of the ⊥s; (2) of the bisectors of the base angles?

17. Of all △s inscribed in a given ⊙, the equilateral △ is a maximum.

18. The square of the third diagonal of a quadrilateral inscribed in a ⊙ is equal to the sum of the squares of tangents to the ⊙ from its extremities.

19. The ⊙, whose diameter is the third diagonal of a quadrilateral inscribed in another ⊙, cuts the latter orthogonally.

20. If from any point in the circumference of a ⊙ three lines be drawn to the angular points of an inscribed equilateral △, one of these lines is equal to the sum of the other two.

21. If the feet of the ⊥ of a △ be joined, the △ thus formed will have its angles bisected by the ⊥s of the original triangle.

22. If all the sides of a quadrilateral or polygon, except one, be given in magnitude and order, the area will be a maximum, when the remaining side is the diameter of a semicircle passing through all the vertices.

23. The area will be the same in whatever order the sides are placed.

24. If two quadrilaterals or polygons have their sides equal, each to each, and if one be inscribed in a ⊙, it will be greater than the other.

25. If from any point P without a ⊙ a secant be drawn cutting the ⊙ in the points A, B; then if C be the middle point of the polar of P, the ∠ ACB is bisected by the polar of P.

26. If OPP′ be any line cutting a ⊙, J, in the points PP′; then if two ⊙s passing through O touch J in the points P, P′, respectively, the difference between their diameters is equal to the diameter of J.

27. Given the base, the difference of the base ∠s, and the sum or difference of the sides of a △, construct it.

28. Given the base, the vertical ∠, and the bisector of the vertical ∠ of a △, construct it.

29. Draw a right line through the point of intersection of two ⊙s, so that the sum or the difference of the squares of the intercepted segments shall be given.

30. If an arc of a ⊙ be divided into two equal, and into two unequal parts, the rectangle contained by the chords of the unequal parts, together with the square of the chord of the arc between the points of section, is equal to the square of the chord of half the arc.

31. If A, B, C, D be four points, ranged in order on a straight line, find on the same line a point O, such that the rectangle OA . OD shall be equal to the rectangle OB . OC.

32. In the same case find the locus of a point P if the ∠ APB equal ∠ CPD.

33. Given two points A, B, and a ⊙ X, find in X a point C, so that the ∠ ACB may be either a maximum or a minimum.

34. The bisectors of the ∠s, at the extremities of the third diagonal of a quadrilateral inscribed in a ⊙ are ⊥ to each other.

35. If the base and the sum of the sides of a △ be given, the rectangle contained by the ⊥s from the extremities of the base on the bisector of the external vertical ∠ is given.

36. If any hexagon be inscribed in a ⊙, the sum of the three alternate ∠s is equal to the sum of the three remaining angles.

37. A line of given length MN slides between two fixed lines OM, ON; then, if MP, NP be ⊥ to OM, ON, the locus of P is a circle.

38. State the theorem corresponding to 35 for the internal bisector of the vertical angle.

39. If AB, AC, AD be two adjacent sides and the diagonal of a ▭, and if a ⊙ passing through A cut these lines in the points P, Q, R, then

$$AB . AP + AC . AQ = AD . AR.$$

40. Draw a chord CD of a semicircle ∥ to a diameter AB, so as to subtend a right ∠ at a given point P in AB (see Exercise 16, Book II.)

41. Find a point in the circumference of a given ⊙, such that the lines joining it to two fixed points in the circumference may make a given intercept on a given chord of the circle.

42. In a given ⊙ describe a △ whose three sides shall pass through three given points.

43. If through any point O three lines be drawn respectively ∥ to the three sides of a △, intersecting the sides in the points A, A′, B, B′, C, C′, then the sum of the rectangles AO . OA′, BO . OB′, CO . OC′ is equal to the rectangle contained by the segments of the chord of the circumscribed ⊙ which passes through O.

44. The lines drawn from the centre of the circle described about a △ to the angular points are ⊥ to the sides of the △ formed by joining the feet of the ⊥s of the original triangle.

45. If a ⊙ touch a semicircle and two ordinates to its diameter, the rectangle under the remote segments of the diameter is equal to the square of the ⊥ from the centre of the ⊙ on the diameter of the semicircle.

46. If AB be the diameter of a semicircle, and AC, BD two chords intersecting in O, the ⊙ about the △ OCD intersects the semicircle orthogonally.

47. If the sum or difference of the tangents from a variable point to two ⊙s be equal to the part of the common tangent of the two ⊙s between the points of contact, the locus of the point is a right line.

48. If pairs of common tangents be drawn to three ⊙s, and if one triad of common tangents be concurrent, the other triad will also be concurrent.

49. The distance between the feet of ⊥s from any point in the circumference of a ⊙ on two fixed radii is equal to the ⊥ from the extremity of either of these radii on the other.

BOOK FOURTH.

SECTION I.

Additional Propositions.

Prop. 1.—*If a circle be inscribed in a triangle, the distances from the angular points of the triangle to the points of contact on the sides are respectively equal to the remainders that are left, when the lengths of the sides are taken separately from their half sum.*

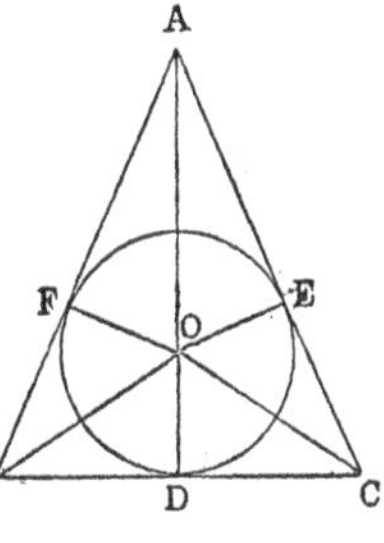

Dem.—Let ABC be the △, D, E, F, the points of contact. Now, since the tangents from an external point are equal, we have AE = AF, BD = BF, CD = CE. Hence AE + BC = AB + CE = half sum of the three sides BC, CA, AB; and denoting these sides by the letters a, b, c, respectively, and half their sum by s, we have

$$AE + a = s$$

therefore $\quad AE = s - a.$

In like manner $\quad BD = s - b; \; CE = s - c.$

Cor. 1.—If r denote the radius of the inscribed ⊙, the area of the triangle $= rs$.

For, let O be the centre of the inscribed ⊙, then we have

$$BC \,.\, r = 2 \triangle BOC,$$
$$CA \,.\, r = 2 \triangle COA,$$
$$AB \,.\, r = 2 \triangle AOB;$$

therefore $\quad (BC + CA + AB)\, r = 2 \triangle ABC;$

that is, $2sr = 2 \triangle ABC$;

therefore $sr = \triangle ABC$. (α)

Cor. 2.—If the ⊙ touch the side BC externally, and the sides AB, AC produced; that is, if it be an escribed ⊙, and if the points of contact be denoted by D′, E′, F′, it may be proved in the same manner that AE′ = AF′ = s; BD′ = BF′ = $s - c$; CD′ = CE′ = $s - b$.

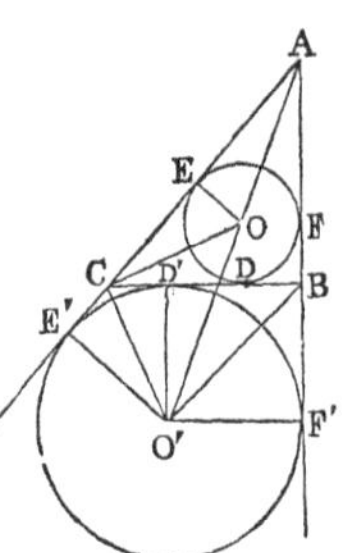

These Propositions, though simple, are very important.

Cor. 3.—If r' denote the radius of the escribed ⊙, which touches the side BC (a) externally,

$$r'(s - a) = \triangle ABC.$$

Dem.— E′O′ . AC = 2 △ AO′C;

that is $r' . b = 2 \triangle AO'C$.

In like manner $r' . c = 2 \triangle AO'B$,

and $r' . a = 2 \triangle BO'C$.

Hence $r'(b + c - a) = 2 \triangle ABC$;

that is $r' . 2(s - a) = 2 \triangle ABC$;

therefore $r' . (s - a) = \triangle ABC$. (β)

Cor. 4.—The rectangle $r . r' = (s - b)(s - c)$.

Dem.—Since CO bisects the ∠ ACB, and CO′ bisects the ∠ BCE′, CO is at right ∠s to CO′; ∴ the ∠ ECO + E′CO′ = a right ∠; and ∠ ECO + COE = one right ∠; ∴ E′CO′ = COE. Hence the △s E′CO′, EOC are equiangular; and, therefore,

E′O′ . EO = E′C . CE; (III. 9.)

that is $r . r' = (s - b)(s - c)$. (γ)

Cor. 5.—If we denote the area of the $\triangle$ ABC by N, we shall have

$$N = \sqrt{s(s-a)(s-b)(s-c)}.$$

For, by equations (α) and (β), we have

$$rs = N, \text{ and } r'(s-a) = N.$$

Therefore, multiplying and substituting from (γ), we get

$$N^2 = s(s-a)(s-b)(s-c);$$

therefore $N = \sqrt{s(s-a)(s-b)(s-c)}.$

Cor. 6.— $N = \sqrt{r \cdot r' \cdot r'' \cdot r'''}$;

where r'', r''' denote the radii of the escribed circles, which touch the sides b, c, externally.

Cor. 7.—If the $\triangle$ ABC be right-angled, having the angle C right,

$$r = s - c; \; r' = s - b; \; r'' = s - a; \; r''' = s.$$

Prop. 2.—*If from any point perpendiculars be let fall on the sides of a regular polygon of n sides, their sum is equal to n times the radius of the inscribed circle.*

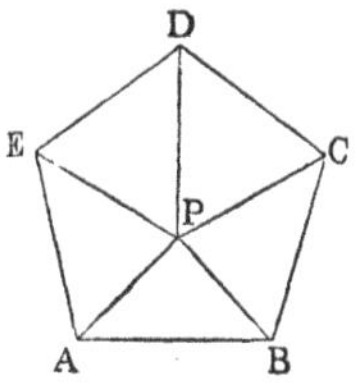

Dem.—Let the given polygon be, say a pentagon ABCDE, and P the given point, and the $\perp$s from P on the sides AB, BC, &c., be denoted by p_1, p_2, p_3, &c., and let the common length of the sides of the polygon be s; then

$$2 \triangle APB = sp_1;$$
$$2 \triangle BPC = sp_2;$$
$$2 \triangle CPD = sp_3;$$
&c., &c.;

therefore, by addition, twice the pentagon

$$= s(p_1 + p_2 + p_3 + p_4 + p_5).$$

Again, if we suppose O to be the centre of the inscribed circle, and R its radius, we get, evidently,

$$2 \triangle AOB = Rs;$$

but the pentagon = 5 △ AOB; therefore

$$\text{twice pentagon} = 5Rs;$$

therefore $s(p_1 + p_2 + p_3 + p_4 + p_5) = 5Rs.$

Hence $p_1 + p_2 + p_3 + p_4 + p_5 = 5R.$

Prop. 3.—*If a regular polygon of n sides be described about a circle, the sum of the perpendiculars from the points of contact on any tangent to the circle equal* nR.

Dem.—Let A, B, C, D, E, &c., be the points of contact of the sides of the polygon with the ⊙, L any tangent to the ⊙, and P its point of contact. Now, the ⊥s from the points A, B, C, &c., on L, are respectively equal to the ⊥s from P on the tangents at the same points; but the sum of the ⊥s from P on the tangents at the points A, B, C, &c., = nR (2). Hence the sum of the ⊥s from the points A, B, C, &c., on L = nR.

Cor. 1.—The sum of the ⊥s from the angular points of an inscribed polygon of n sides upon any line equal n times the ⊥ from the centre on the same line.

Cor. 2.—The centre of mean position of the angular points of a regular polygon is the centre of its circumscribed circle.

For, since there are n points, the sum of the ⊥s from these points on any line equal n times the ⊥ from their centre of mean position on the line (I., 17); therefore the ⊥ from the centre of the circumscribed ⊙ on any line is equal to the ⊥ from the centre of mean position on the same line; and, consequently, these centres must coincide.

Cor. 3.—The sum of the ⊥s from the angular points of an inscribed polygon on any diameter is zero; or, in other words, the sum of the ⊥s on one side of the diameter is equal to the sum of the ⊥s on the other side.

Prop. 4.—*If a regular polygon of n sides be inscribed in a circle, whose radius is* R, *and if* P *be any point whose distance from the centre of the circle is* R′, *then the sum of the squares of all the lines from* P *to the angular points of the polygon is equal to* $n(R^2 + R'^2)$.

Dem.—Let O be the centre of the ⊙, then O is the mean centre of the angular points; hence (II., 10) the sum of the squares of the lines drawn from P to the angular points exceeds the sum of the squares of the lines drawn from O by $n\mathrm{OP}^2$, that is by $n\mathrm{R}'^2$; but all the lines drawn from O to the angular points are equal to one another, each being the radius. Hence the sum of their squares is $n\mathrm{R}^2$. Hence the Proposition is proved.

Cor. 1.—If the point P be in the circumference of the ⊙, we have the following theorem:—*The sum of the squares of the lines drawn from any point in the circumference of a circle to the angular points of an inscribed polygon is equal to* $2n\mathrm{R}^2$.

The following is an independent proof of this theorem:—It is seen at once, if we denote the ⊥s from the angular points on the tangent at P by p_1, p_2, &c., that

$$2\mathrm{R} \,.\, p_1 = \mathrm{AP}^2;$$
$$2\mathrm{R} \,.\, p_2 = \mathrm{BP}^2;$$
$$2\mathrm{R} \,.\, p_3 = \mathrm{CP}^2, \text{ \&c.}$$

Hence

$$2\mathrm{R}\,(p_1 + p_2 + p_3 + \text{\&c.}) = \mathrm{AP}^2 + \mathrm{BP}^2 + \mathrm{CP}^2, \text{ \&c.};$$

or

$$2\mathrm{R} \,.\, n\mathrm{R} = \mathrm{AP}^2 + \mathrm{BP}^2 + \mathrm{CP}^2, \text{ \&c.};$$

therefore the sum of the squares of all the lines from $\mathrm{P} = 2n\mathrm{R}^2$.

Cor. 2.—The sum of the squares of all the lines of connexion of the angular points of a regular polygon of n sides, inscribed in a ⊙ whose radius is R, is $n^2\mathrm{R}^2$.

This follows from supposing the point P to coincide with each angular point in succession, and adding all the results, and taking half, because each line occurs twice.

Prop. 5.—*If* O *be the point of intersection of the three perpendiculars* AD, BE, CF *of a triangle* ABC, *and if* G, H, I *be the middle points of the sides of the triangle, and* K, L, M *the middle points of the lines* OA, OB, OC; *then the nine points* D, E, F; G, H, I; K, L, M, *are in the circumference of a circle.*

Dem.—Join HK, HG, IK, IG; then, because AO is bisected in K, and AC in H, HK is ∥ to CO. In like manner, HG is ∥ to AB. Hence the ∠ GHK is equal to the ∠ between CO and AB; ∴ it is a right ∠; consequently, the ⊙ described on GK as diameter passes through H. In like manner, it passes through I; and since the ∠ KDG is right, it passes through D; ∴ the

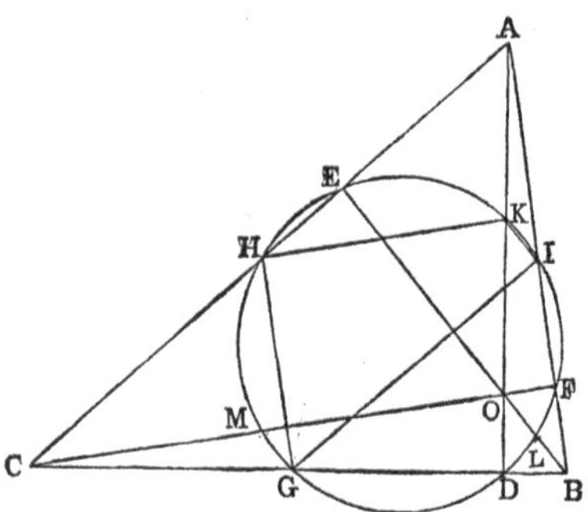

circle through the three points G, H, I, passes through the two points D, K. In like manner, it may be proved that it passes through the pairs of points E, L; F, M. Hence it passes through the nine points.

Def.—*The circle through the middle points of the sides of a triangle is called, on account of the property we have just proved, "The Nine-points Circle of the Triangle."*

Prop. 6.—*To draw the fourth common tangent to the two escribed circles of a plane triangle, which touch the base produced, without describing those circles.*

Con.—From B, one of the extremities of the base, let fall a ⊥ BG on the external bisector AI of the vertical ∠ of the △ ABC; produce BG and AI to meet the sides CA, CB of the △ in the points H and I; then the line joining the points H and I is the fourth common tangent.

Dem.—The △s BGA, HGA have the side AG common, and the ∠s adjacent to this side in the two △s equal each to each; hence AH = AB. Again the △s AHI, ABI have the sides AH, AI and the included ∠ in the one equal to the two sides AB, AI and the included ∠ in the other; ∴ the ∠ HIA = BIA.

Now, bisect the ∠ ABI by the line BO, and it is evident, by letting fall ⊥s on the four sides of the quadrilateral ABIH from the point O, that the four ⊥s are equal to one another. Hence the ⊙, having O as centre, and any of these ⊥s as radius, will be inscribed in the quadrilateral; ∴ HI is a tangent to the escribed ⊙, which touches AB externally. In like manner, it may be proved that HI touches the escribed ⊙, which touches AC externally. Hence HI is the fourth common tangent to these two circles.

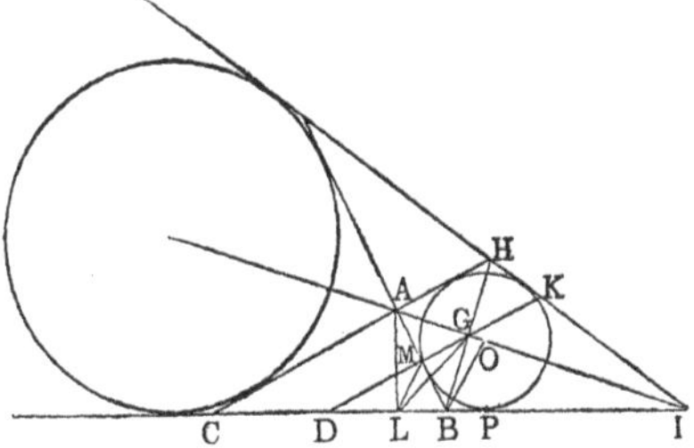

Cor. 1.—If D be the middle point of the base BC, the ⊙, whose centre is D and whose radius is DG, is orthogonal to the two escribed ⊙s which touch BC produced.

For, let P be the point of contact of the escribed ⊙, which touches AB externally, then

$$PD = CP - CD = \tfrac{1}{2}(a + b + c) - \tfrac{1}{2}a = \tfrac{1}{2}(b + c);$$

and since BH is bisected in G, and BC in D,

$$DG = \tfrac{1}{2}CH = \tfrac{1}{2}(AB + AC) = \tfrac{1}{2}(b + c);$$

therefore the ⊙, whose centre is D and radius DG, will cut orthogonally the ⊙ which touches at P.

Cor. 2.—Let DG cut AB in M, and HI in K, and from A let fall the ⊥ AL, then the quadrilateral LMKI is inscribed in a circle.

For, since the ∠s ALB, AGB are right, ALBG is a quadrilateral in a ⊙, and M is the centre of the ⊙; ∴ ML = MB, and ∠ MLB = MBL. Again, ∠ MKI = AHI = ABI; ∴ MKI + MLI = ABI + MBL = two right ∠s. Hence MKIL is a quadrilateral inscribed in a circle.

Prop. 7.—*The "Nine-points Circle" is the inverse of the fourth common tangent to the two escribed circles which touch the base produced, with respect to the circle whose centre is at the middle point of the base, and which cuts these circles orthogonally.*

Dem.—The ∠ DML (see fig., last Prop.) = twice DGL (III. xx.); and the ∠ HIL = twice AIL; but DML = HIL, since MKIL is a quadrilateral in a ⊙; ∴ the ∠ DGL = GIL. Hence, if a ⊙ be described about the △ GIL it will touch the line GD (III. xxxii.); ∴ DL . DI = DG^2; ∴ the point L is the inverse of the point I, with respect to the ⊙ whose centre is D and radius DG. Again, since MKIL is a quadrilateral in a ⊙, DM . DK = DL . DI, and, ∴ = DG^2. Hence the point M is the inverse of K, and ∴ the ⊙ described through the points DLM is the inverse of the line HI (III. 20); that is, the "Nine-points Circle" is the inverse of the fourth common tangent, with respect to the ⊙ whose centre is the middle point of the base, and whose radius is equal to half the sum of the two remaining sides.

Cor. 1.—In like manner, it may be proved that the "Nine-points Circle" is the inverse of the fourth common tangent to the inscribed ⊙ and the escribed ⊙, which touches the base externally, with respect to the ⊙ whose centre is the middle point of the base, and whose radius is = to half the difference of the remaining sides.

Cor. 2.—The "Nine-points Circle" touches the inscribed and the escribed circles of the triangle.

For, since it is the inverse of the fourth common

tangent to the two escribed ⊙s which touch the base produced, with respect to the ⊙ whose centre is D, and which cuts these ⊙s orthogonally; if we join D to the points of contact of the fourth common tangent, the points where the joining lines meet these ⊙s again will be the inverses of the points of contact. Hence they will be common both to the "Nine-points Circle" and the escribed ⊙s; ∴ the "Nine-points Circle" touches these escribed ⊙s in these points; and in a similar way the points of contact with the inscribed ⊙ and the escribed ⊙ which touch the base externally may be found.

Cor. 3.—Since the "Nine-points Circle" of a plane △ is also the "Nine-points Circle" of each of the three △s into which it is divided by the lines drawn from the intersection of its ⊥s to the angular points, we see that the "Nine-points Circle" touches also the inscribed and escribed circles of each of these triangles.

Prop. 8.—The following Propositions, in connexion with the circle described about a triangle, are very important:—

(1). *The lines which join the extremities of the diameter, which is perpendicular to the base of a triangle, to the vertical angle, are the internal and external bisectors of the vertical angle.*

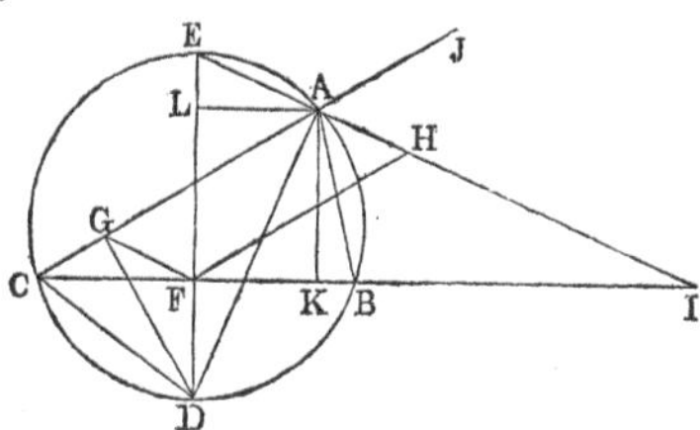

Dem.—Let DE be the diameter ⊥ to BC. Join AD, AE. Produce AE to meet CB in I. Now, from the construction, we have the arc CD = the arc BD. Hence the ∠ CAD = DAB; ∴ AD is the internal bisector of the ∠ CAB. Again, since DE is the diameter of the ⊙,

the ∠ DAE is right; ∴ the ∠ DAE = DAH; and from these, taking away the equal ∠s CAD, DAB, we have the ∠ CAE = BAH; ∴ JAH = BAH. Hence AH is the external bisector.

(2). *If from* D *a perpendicular be let fall on* AC, *the segments* AG, GC *into which it divides* AC *are respectively the half sum and the half difference of the sides* AB, AC.

Dem.—Join CD, GF. Draw FH ∥ to AC. Since the ∠s CGD, CFD are right, the figure CGFD is a quadrilateral in a ⊙. Hence the ∠ AGF = CDE (III., xxii.) = CAE (III., xxi.); ∴ GF is ∥ to AE. Hence AHFG is a ▭; and AG = FH = $\frac{1}{2}$ sum of AB, AC (I., 11, *Cor.* 1). Again, GC = AC − AG = AC − $\frac{1}{2}$(AB + AC) = $\frac{1}{2}$(AC − AB).

(3). *If from* E *a perpendicular* EG′ *be drawn to* AC, CG′ *and* AG′ *are respectively the half sum and the half difference of* AC, AB.

This may be proved like the last.

(4). *Through* A *draw* AL *perpendicular to* DE. *The rectangle* DL . EF *is equal to the square of half the sum of the sides* AC, AB.

Dem.—The △s ALD, EFI have evidently the ∠s at D and I equal, and the right ∠s at L and F are equal. Hence the △s are equiangular; ∴ DL . EF = AL . FI = FK . FI = the square of half the sum of the sides (Prop. 7).

(5). In like manner it may be proved that EL . FD *is equal to the square of half the difference of* AC, AB.

Prop. 9.—*If* a, b, c *denote, as in* Prop. 1, *the lengths of the sides of the triangle* ABC, *then the centre of the inscribed circle will be the centre of mean position of its angular points for the system of multiples* a, b, c.

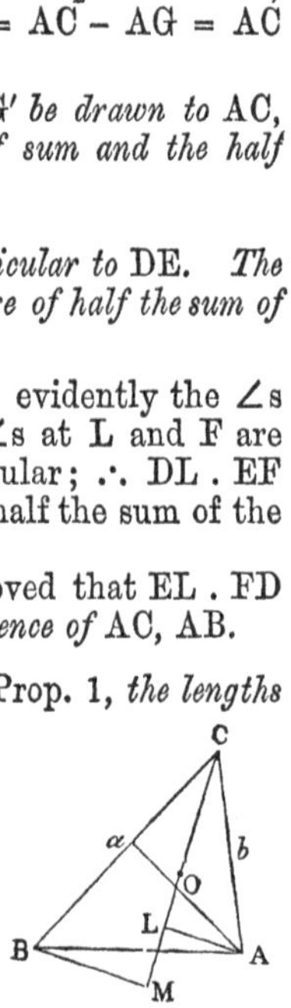

Dem.—Let O be the centre of the inscribed ⊙. Join CO; and on CO produced let fall the ⊥s AL, BM. Now, the △s ACL, BCM have the ∠ ACL = BCM;

and the $\angle$ ALC = BMC. Hence they are equiangular

therefore BC . AL = AC . BM; (III. 9)

or a . AL = b . BM. (α)

Now, if we introduce the signs + and −, since the ⊥s AL, BM fall on different sides of CL, they must be affected with contrary signs; ∴ the equation (α) expresses that a times the ⊥ from A on CO + b times the ⊥ from B on CO = 0; and since the ⊥ from C on CO is evidently = 0, we have the sum of a times perpendicular from A; b times perpendicular from B; c times perpendicular from C, on the line CO = 0. Hence the line CO passes through the centre of mean position for the system of multiples a, b, c. In like manner, AO passes through the centre of mean position. And since a point which lies on each of two lines must be their point of intersection, O must be the centre of mean position for the system of multiples a, b, c.

Cor. 1.—If O′, O″, O‴ be the centres of the escribed ⊙s, O′ is the centre of mean position for the system of multiples $-a$, $+b$, $+c$; O″ for the system $+a$, $-b$, $+c$: and O‴ for the system $+a$, $+b$, $-c$.

SECTION II.

Exercises.

1. The square of the side of an equilateral △ inscribed in a ⊙ equal three times the square of the radius.

2. The square described about a ⊙ equal twice the inscribed square.

3. The inscribed hexagon equal twice the inscribed equilateral triangle.

4. In the construction of IV., x., if F be the second point in which the ⊙ ACD intersects the ⊙ BDE, and if we join AF, DF, the △ ADF has each of its base ∠s double the vertical ∠. The same property holds for the △s ACF, BCD.

5. The square of the side of a hexagon inscribed in a ⊙, together with the square of the side of a decagon, is equal to the square of the side of a pentagon.

6. Any diagonal of a pentagon is divided by a consecutive diagonal into two parts, such that the rectangle contained by the whole and one part is equal to the square of the other part.

7. Divide an ∠ of an equilateral △ into five equal parts.

8. Inscribe a ⊙ in a given sector of a circle.

9. The locus of the centre of the ⊙ inscribed in a △, whose base and vertical ∠ are given, is a circle.

10. If tangents be drawn to a ⊙ at the angular points of an inscribed regular polygon of any number of sides, they will form a circumscribed regular polygon.

11. The line joining the centres of the inscribed and circumscribed ⊙s subtends at any of the angular points of a △ an ∠ equal to half the difference of the remaining angles.

12. Inscribe an equilateral △ in a given square.

13. The six lines of connexion of the centres of the inscribed and escribed ⊙s of a plane △ are bisected by the circumference of the circumscribed circle.

14. Describe a regular octagon in a given square.

15. A regular polygon of any number of sides has one ⊙ inscribed in it, and another circumscribed about it, and the two ⊙s are concentric.

16. If O, O′, O″, O‴, be the centres of the inscribed and escribed ⊙s of a plane △, then O is the mean centre of the points O′, O″, O‴, for the system of multiples $(s-a)$, $(s-b)$, $(s-c)$.

17. In the same case, O′ is the mean centre of the points O, O″, O‴, for the system of multiples s, $s-b$, $s-c$, and corresponding properties hold for the points O″, O‴.

18. If r be the radius of the ⊙ inscribed in a △, and ρ_1, ρ_2 the radii of two ⊙s touching the circumscribed ⊙, and also touching each other at the centre of the inscribed ⊙; then

$$\frac{2}{r}=\frac{1}{\rho_1}+\frac{1}{\rho_2}.$$

19. If r, r_1, r_2, r_3 be the radii of the inscribed and escribed ⊙s of a plane △, and R the radius of the circumscribed ⊙; then

$$r_1+r_2+r_3-r=4R.$$

20. In the same case,

$$\frac{1}{r}=\frac{1}{r_1}+\frac{1}{r_2}+\frac{1}{r_3}.$$

21. In a given ⊙ inscribe a △, so that two of its sides may pass through given points, and that the third side may be a maximum.

22. What theorem analogous to 18 holds for escribed ⊙s?

23. Draw from the vertical ∠ of an obtuse-angled △ a line to a point in the base, such that its square will be equal to the rectangle contained by the segments of the base.

24. If the line AD, bisecting the vertical ∠ A of the △ ABC, meets the base BC in D, and the circumscribed ⊙ in E, then the line CE is a tangent to the ⊙ described about the △ ADC.

25. The sum of the squares of the ⊥s from the angular points of a regular polygon inscribed in a ⊙ upon any diameter of the ⊙ is equal to half n times the square of the radius.

26. Given the base and vertical ∠ of a △, find the locus of the centre of the ⊙ which passes through the centres of the three escribed circles.

27. If a ⊙ touch the arcs AC, BC, and the line AB in the construction of Euclid (I. i.), prove its radius equal to $\frac{3}{8}$ of AB.

28. Given the base and the vertical ∠ of a △, find the locus of the centre of its "Nine-point Circle."

29. If from any point in the circumference of a ⊙ ⊥s be let fall on the sides of a circumscribed regular polygon, the sum of their squares is equal to $\frac{3}{2}n$ times the square of the radius.

30. The internal and external bisectors of the ∠s of the △, formed by joining the middle points of the sides of another △, are the six radical axes of the inscribed and escribed ⊙s of the latter.

31. The ⊙ described about a △ touches the sixteen circles inscribed and escribed to the four △s formed by joining the centres of the inscribed and escribed circles of the original triangle.

32. If O, O′ have the same meaning as in question 16, then

$$AO \,.\, AO' = AB \,.\, AC.$$

34. Given the base and the vertical ∠ of a △, find the locus of the centre of a ⊙ passing through the centre of the inscribed circle, and the centres of any two escribed circles.

BOOK SIXTH.

SECTION I.

Additional Propositions.

Prop. 1.—*If two triangles have a common base, but different vertices, they are to one another as the segments into which the line joining the vertices is divided by the common base or base produced.*

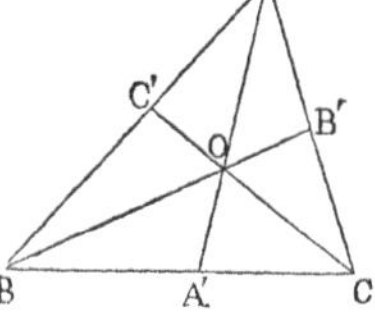

Let the two △s be AOB, AOC, having the base AO common; let AO cut the line BC, joining the vertices in A′; then

$$\text{AOB} : \text{AOC} :: \text{BA}' : \text{A}'\text{C}.$$

Dem.—The △ $\text{ABA}' : \text{ACA}' :: \text{BA}' : \text{A}'\text{C}$;

and $$\text{OBA}' : \text{OCA}' :: \text{BA}' : \text{A}'\text{C};$$

therefore

$$\text{ABA}' - \text{OBA}' : \text{ACA}' - \text{OCA}' :: \text{BA}' : \text{A}'\text{C};$$

or $$\text{AOB} : \text{AOC} :: \text{BA}' : \text{A}'\text{C}.$$

Prop. 2.—*If three concurrent lines* AO, BO, CO, *drawn from the angular points of a triangle, meet the opposite sides in the points* A′, B′, C′, *the product of the three ratios*

$$\frac{\text{BA}'}{\text{A}'\text{C}},\ \frac{\text{CB}'}{\text{B}'\text{A}},\ \frac{\text{AC}'}{\text{C}'\text{B}}\ \textit{is unity.}$$

Dem.—From the last Proposition, we have

$$\frac{BA'}{A'C} = \frac{AOB}{AOC};$$

$$\frac{CB'}{B'A} = \frac{BOC}{BOA};$$

$$\frac{AC'}{C'B} = \frac{AOC}{BOC}.$$

Hence, multiplying out, we get the product equal to unity.

Cor. This may be written

$$AB' . BC' . CA' = A'B . B'C . C'A.$$

The symmetry of this expression is apparent. Expressed in words, it gives the product of three alternate segments of the sides equal to the product of the three remaining segments.

Prop. 3.—*If two parallel lines be intersected by three concurrent transversals, the segments intercepted by the transversals on the parallels are proportional.*

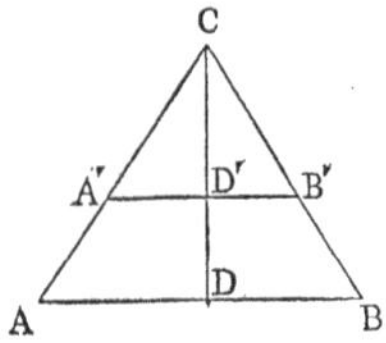

Let the ||s be AB, A'B', and the transversals CA, CD, CB; then

$$AD : DB :: A'D' : D'B'.$$

Dem. — The triangles ADC, A'D'C are equiangular;

therefore $AD : DC :: A'D' : D'C$.

In like manner, $DC : DB :: D'C : D'B'$;

therefore *ex aequali* $AD : DB :: A'D' : D'B'$.

Cor.—If from the points D, D' we draw two ⊥s DE, D'E' to AC, and two ⊥s DF, D'F' to BC; then

$$DE : DF :: D'E' : D'F'.$$

Prop. 4.—*If the sides of a triangle* ABC *be cut by any transversal, in the points* A′, B′, C′; *then the product of the three ratios*

$$\frac{AB'}{B'C'}, \frac{BC'}{C'A'}, \frac{CA'}{A'B}$$

is equal to unity.

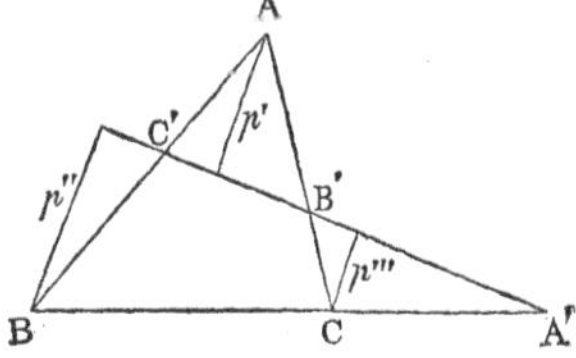

Dem. — From the points A, B, C let fall the ⊥s p', p'', p''' on the transversal; then, by similar △s the three ratios are respectively $= \frac{p'}{p'''}, \frac{p''}{p'}, \frac{p'''}{p''}$, and the product of these is evidently equal unity. Hence the proposition is proved.

Observation.—If we introduce the signs plus and minus, in this Proposition, it is evident that one of the three ratios must be negative. And when the transversal cuts all the sides of the triangle externally, all three will be negative. Hence their product will, in all cases, be equal to negative unity.

Cor. 1.—If A′, B′, C′ be three points on the sides of a triangle, either all external, or two internal and one external, such that the product of the three ratios

$$\frac{AB'}{B'C'}, \frac{BC'}{C'A'}, \frac{CA'}{A'B}$$

is equal to negative unity, then the three points are collinear.

Cor. 2.—The three external bisectors of the angles of a triangle meet the sides in three points, which are collinear.

For, let the meeting points be A′, B′, C′, and we have the ratios

$$\frac{BA'}{A'C'}, \frac{CB'}{B'A'}, \frac{AC'}{CB'} = \text{to the ratios } \frac{BA}{AC'}, \frac{CB}{BA'}, \frac{AC}{CB'}$$

respectively; and, therefore, their produce is unity.

Prop. 5.—*In any triangle, the rectangle contained by two sides is equal to the rectangle contained by the perpendicular on the third side and the diameter of the circumscribed circle.*

Let ABC be the △, AD the ⊥, AE the diameter of the ⊙; then AB . AC = AE . AD.

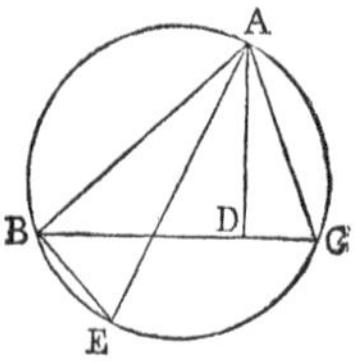

Dem.—Since AE is the diameter, the ∠ ABE is right, and ADC is right; ∴ ABE = ADC; and AEB = ACD (III., xxi.); therefore the △s ABE and ADC are equiangular; and AB : AE : : AD : AC (iv.). Hence AB . AC = AE . AD.

Cor.—If a, b, c denote the three sides of a triangle, and R the radius of the circumscribed circle, then the area of the triangle $= \dfrac{abc}{4R}$.

For, let AD be denoted by p, we have (5)

$$2pR = bc;$$

therefore $$2apR = abc,$$

$$\frac{ap}{2} = \frac{abc}{4R};$$

that is, area of triangle $= \dfrac{abc}{4R}$.

Prop. 6.—*If a figure of any even number of sides be inscribed in a circle, the continued product of the perpendiculars let fall from any point in the circumference on the odd sides is equal to the continued product of the perpendiculars on the even sides.*

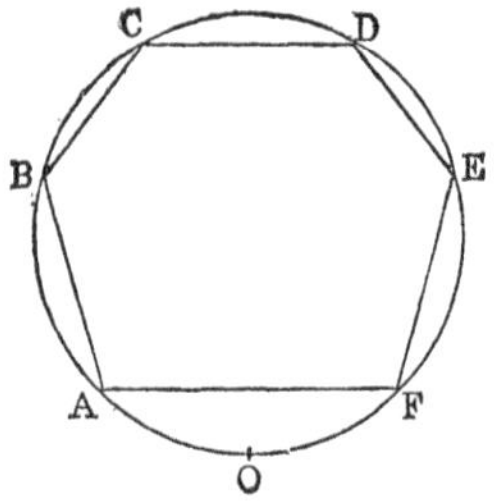

We shall prove this Proposition for the case of a hexagon, and then it will be evident that the proof is general.

Let ABCDEF be the hexagon, O the point, and let the ⊥s from O on the lines AB, BC . . . FA, be denoted by α, β, γ, δ, ϵ, ϕ; let D denote the diameter of the ⊙, and let the lengths of the six lines OA, OB . . . OF be denoted by l, m, n, p, q, r; then we have $D\alpha = lm$; $D\gamma = np$; $D\epsilon = qr$;

therefore $D^3\alpha\gamma\epsilon = lmnpqr$.

In like manner, $D^3\beta\delta\phi = lmnpqr$;

therefore $\alpha\gamma\epsilon = \beta\delta\phi$. (Q.E.D.)

Cor. 1.—The six points A, B, C, D, E, F may be taken in any order of sequence, and the Proposition will hold; or, in other words, if we draw all the diagonals of the hexagon, and take any three lines, such as AC, BD, EF, which terminate in the six points A, B, C, D, E, F, then the continuous product of the ⊥s on them will be equal to the continuous product of the ⊥s on any other three lines also terminating in the six points.

Cor. 2.—When the figure inscribed in the circle contains only four sides, this Proposition is the theorem proved (III., 11.)

Cor. 3.—If we suppose two of the angular points to become infinitely near; then the line joining these points, if produced, will become a tangent to the circle, and we shall in this way have a theorem that will be true for a polygon of an odd number of sides.

Cor. 4.—If perpendiculars be let fall from any point in the circumference of a circle on the sides of an inscribed triangle, their continued product is equal to the continued product of the perpendiculars from the same point on the tangents to the circle at the angular points.

Prop. 7.—*Given, in magnitude and position, the base* BC *of a triangle and the ratio* BA : AC *of the sides, it is required to find the locus of its vertex* A.

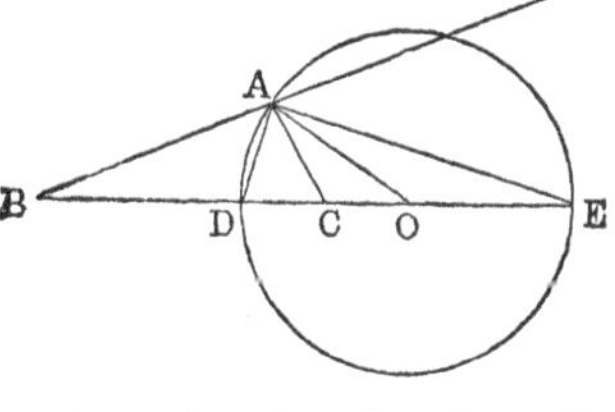

Bisect the internal and the external vertical angles by the lines AD, AE. Now, BA : AC : : BD : DC (III.); but the ratio BA : AC is given (Hyp.); therefore the ratio BD : DC is given, and BC is given (Hyp.); ∴ the point D is given. In like manner the point E is given. Again, the angle DAE is evidently equal half the sum of the angles BAC, CAF. Hence DAE is right, and the circle described on the line DE as diameter will pass through A, and will be the required locus.

Cor. 1.—The circle described about the triangle ABC will cut the circle DAE orthogonally.

For, let O be the centre of the ⊙ DAE. Join AO; then the angle DAO = ADO, that is, DAC + CAO = BAD + ABO; but BAD = DAC; ∴ CAO = ABO; ∴ AO touches the ⊙ described about the △ BAC. Hence the ⊙s cut orthogonally.

Cor. 2.—Any circle passing through the points B, C, is cut orthogonally by the circle DAE.

Cor. 3.—If we consider each side of the triangle as base in succession, the three circles which are the loci of the vertices have two points common.

Prop. 8.—*If through* O, *the intersection of the diagonals of a quadrilateral* ABCD, *a line* OH *be drawn parallel to one of the sides* AB, *meeting the opposite side* CD *in* G, *and the third diagonal in* H, OH *is bisected in* G.

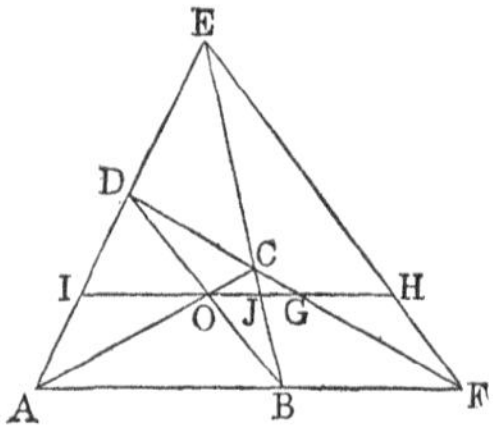

Dem.—Produce HO to meet AD in I, and let it meet BC in J.

Now IJ : JH : : AB : BF, (Prop. 3.)

and OJ : JG : : AB : BF;

therefore IO : GH : : AB : BF;

but AB : BF : : IO : OG; ∴ OG = GH.

Cor.—GO is a mean proportional between GJ and GI.

Prop. 9.—*If a triangle given in species have one angular point fixed, and if a second angular point moves along a given line, the third will also move along a given line.*

Let ABC be the △ which is given in species; let the point A be fixed; the point B move along a given line BD: it is required to find the locus of C.

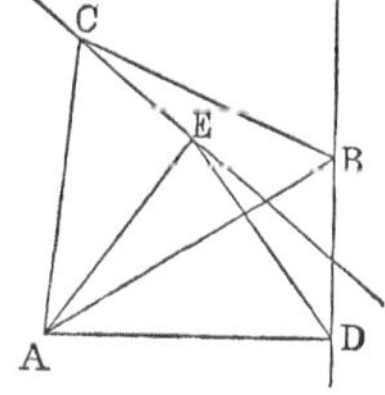

From A let fall the ⊥ AD on BD; on AD describe a △ ADE equiangular to the △ ABC; then the △ ADE is given in position; ∴ E is a given point. Join EC. Now, since the △s ADE, ABC are equiangular, we have

$$AD : AE :: AB : AC;$$

therefore $$AD : AB :: AE : AC:$$

and the angle DAB is evidently = EAC. Hence the △s DAB, EAC are equiangular; ∴ the angle ADB = AEC. Hence the angle AEC is right, and the line EC is given in position; ∴ the locus of C is a right line.

Cor.—By an obvious modification of the foregoing demonstration we can prove the following theorem:—If a △ be given in species, and have one angular point given in position; then if a second angular point move along a given ⊙, the locus of the third angular point is a circle.

Prop. 10.—*If* O *be the centre of the inscribed circle of the triangle* ABC, then $AO^2 : AB . AC :: s - a : s$.

Dem.—Let O′ be the centre of the escribed ⊙ touching BC externally; let fall the ⊥s OD, O′E. Join OB, OC, O′B, O′C. Now, the ∠s O′BO, O′CO are evidently right ∠s; ∴ OBO′C is a quadrilateral inscribed in a circle, and ∠ BO′O = BCO = ACO; and BAO′ = OAC. Hence the triangles O′BA and COA are equiangular; ∴ O′A: BA :: AC : AO; ∴ O′A . OA = AB . AC. Hence

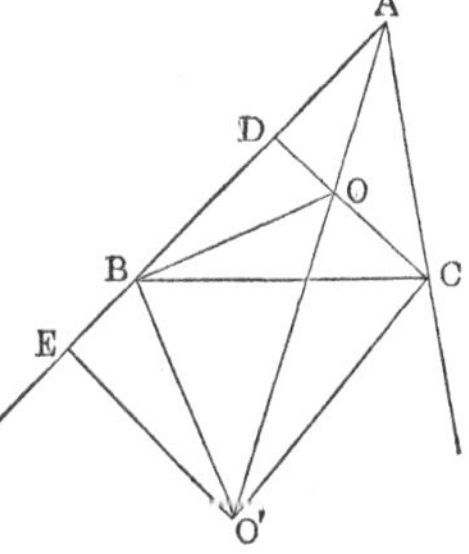

OA² : AB . AC : : OA² : O′A . OA : : OA : O′A : : AD : AE; but AD = $s - a$, and AE = s;

therefore OA² : AB . AC : : $(s - a) : s$.

Cor. 1.— $$\frac{OA^2}{bc} + \frac{OB^2}{ca} + \frac{OC^2}{ab} = 1.$$

For $$\frac{OA^2}{bc} = \frac{s - a}{s}.$$

In like manner, $$\frac{OB^2}{ca} = \frac{s - b}{s},$$

and $$\frac{OC^2}{ab} = \frac{s - c}{s};$$

therefore, by addition,

$$\frac{OA^2}{bc} + \frac{OB^2}{ca} + \frac{OC^2}{ab} = 1.$$

Cor. 2.—If O′, O″, O‴ be the centres of the escribed circles,

$$\frac{O'B^2}{ca} + \frac{O'C^2}{ab} - \frac{O'A^2}{bc} = -1, \text{ \&c.}$$

Prop. 11.—*If r, R be the radii of the inscribed and circumscribed circles of a plane triangle, δ the distance between their centres; then*

$$\frac{r}{R + \delta} + \frac{r}{R - \delta} = 1.$$

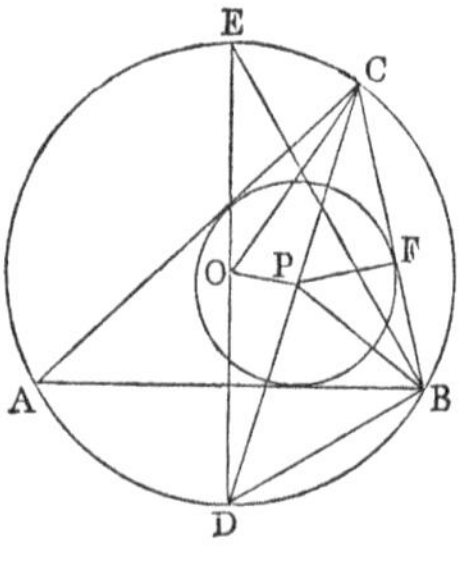

Dem.—Let O, P be the centres of the ⊙s. Join CP, and let it meet the circumscribed ⊙ in D. Join DO, and produce to meet the circumscribed ⊙ in E. Join EB, OP, PF, PB, BD. Since P is the centre of the inscribed ⊙, CP bisects the ∠ ACB; ∴ the arc AD = the arc DB. Hence the ∠ ABD = DCB (III., 21); and because PB bisects the ∠ ABC, the ∠ PBA = PBC; ∴ the ∠ PBD = PCB + PBC = DPB; ∴ DP = DB.

Again, the $\triangle$s DEB, PCF are equiangular; because the angles DEB and PCF are equal, being in the same segment, and the angles DBE and PFC are right. Hence DE : DB : : CP : PF (iv.); $\therefore$ DE . PF = DB . PC = DP . PC.

Now, since the triangle OCD is isosceles, DP . PC $= \mathrm{OC}^2 - \mathrm{OP}^2$ (II., I.);

therefore $\qquad \mathrm{DE} \cdot \mathrm{PF} = \mathrm{OC}^2 - \mathrm{OP}^2$;

that is, $\qquad 2Rr = R^2 - \delta^2$;

therefore $$\frac{r}{R-\delta} + \frac{r}{R+\delta} = 1.$$

Cor. 1.—If r', r'', r''' denote the radii of the escribed $\odot$s, and δ', δ'', δ''' the distances of their centres from the centre of the circumscribed $\odot$, we get in like manner

$$\frac{r'}{R-\delta'} + \frac{r'}{R+\delta'} = -1, \text{ \&c.}$$

Cor. 2.—If O′T′, O″T″, O‴T‴ be the tangents from the points O′, O″, O‴ to the circumscribed $\odot$; then

$$2Rr' = \mathrm{O'T'}^2, \text{ \&c.}$$

Cor. 3.—If through O we describe a $\odot$, touching the circumscribed $\odot$, and touching the diameter of it, which passes through P, this $\odot$ will be equal to the inscribed $\odot$; and similar Propositions hold for circles passing through the points O′, O″, O‴.

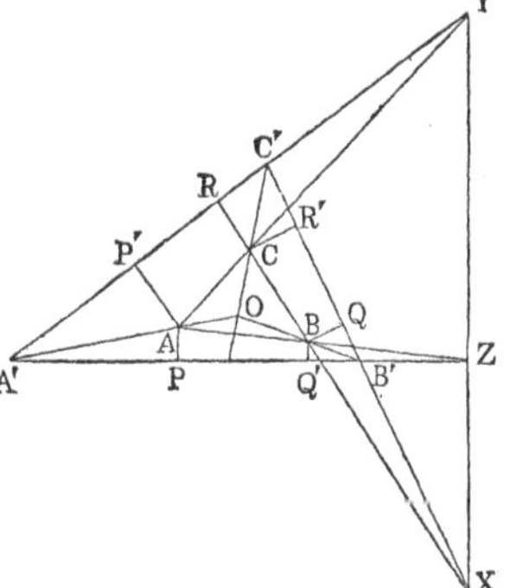

Prop. 12.—*If two triangles be such that the lines joining corresponding vertices are concurrent, then the points of intersection of the corresponding sides are collinear.*

Let ABC, A′B′C′ be the two △s, having the lines joining their corresponding vertices meeting in a point O: it is required to prove that the three points X, Y, Z, which are the intersections of corresponding sides, are collinear.

Dem.—From A, B, C let fall three pairs of ⊥s on the sides of the △ A′B′C′; and from O let fall three ⊥s p', p'', p''' on the sides B′C′, C′A′, A′B′.

Now we have, from *Cor.*, Prop. 3,

$$\frac{AP}{AP'} = \frac{p'''}{p''}, \quad \frac{BQ}{BQ'} = \frac{p'}{p'''}, \quad \frac{CR}{CR'} = \frac{p''}{p'}.$$

Hence the product of the ratios,

$$\frac{AP}{AP'}, \quad \frac{BQ}{BQ'}, \quad \frac{CR}{CR'} = \text{unity}.$$

Again we have, independent of sign, (IV.)

$$\frac{AZ}{ZB} = \frac{AP}{BQ'}, \quad \frac{BX}{XC} = \frac{BQ}{CR'}, \quad \frac{CY}{AY} = \frac{CR}{AP'}.$$

Hence the product of the three ratios

$$\frac{AZ}{ZB}, \quad \frac{BX}{XC}, \quad \frac{CY}{YA}$$

is equal to the product of the three ratios

$$\frac{AP}{BQ'}, \quad \frac{BQ}{CR'}, \quad \frac{CR}{AP'};$$

and, therefore, equal to unity. Hence, by *Cor.*, Prop. 4, the points X, Y, Z are collinear.

Cor.—If two △s be such that the points of intersection of corresponding sides are collinear, then the lines joining corresponding vertices are concurrent.

Observation.—Triangles whose corresponding vertices lie on concurrent lines have received different names from geometers. SALMON and PONCELET call such triangles *homologous*. These writers call the point O the *centre of homology*; and the line XYZ the *axis of homology*. TOWNSEND and CLEBSCH call them triangles in *perspective*; and the point O, and the line XYZ the *centre* and the *axis* of *perspective*.

Prop. 13.—*When three triangles are two by two in perspective, and have the same axis of perspective, their three centres of perspective are collinear.*

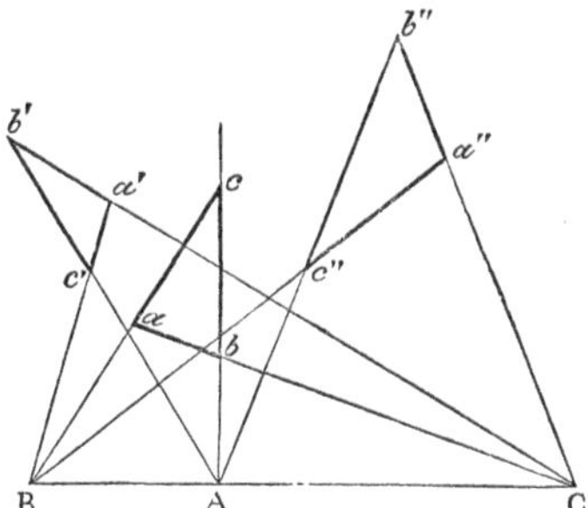

Let abc, $a'b'c'$, $a''b''c''$ be the three △s whose corresponding sides are concurrent in the collinear points A, B, C. Now let us consider the two △s $aa'a''$, $bb'b''$, formed by joining the corresponding vertices a, a', a'', b, b', b'', and we see that the lines ab, $a'b'$, $a''b''$ joining corresponding vertices are concurrent, their centre of perspective being C. Hence the intersections of their corresponding sides are collinear; but the intersections of the corresponding sides of these △s are the centres of perspective of the △s abc, $a'b'c'$, $a''b''c''$. Hence the Proposition is proved.

Cor.—The three △s $aa'a''$, $bb'b''$, $cc'c''$ have the same axis of perspective; and their centres of perspective are the points A, B, C. Hence the centres of perspective of this triad of △s lie on the axis of perspective of the system abc, $a'b'c'$, $a''b''c''$, and conversely.

Prop. 14.—*When three triangles which are two by two in perspective have the same centre of homology, their three axes of homology are concurrent.*

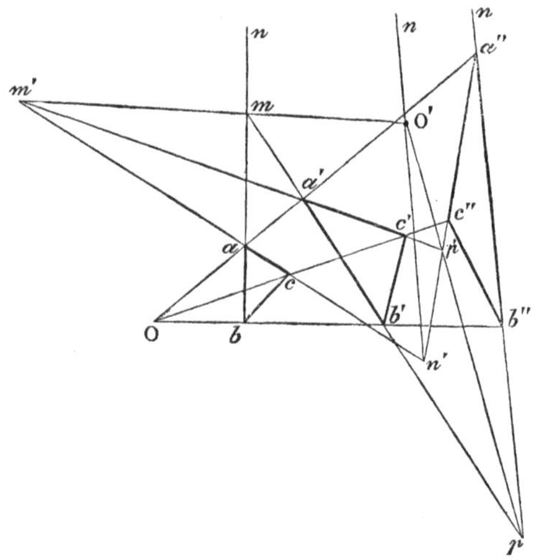

Let abc, $a'b'c'$, $a''b''c''$ be three $\triangle$s, having the point O as a common centre of perspective. Now, let us consider the two $\triangle$s formed by the two systems of lines ab, $a'b'$, $a''b''$; and ac, $a'c'$, $a''c''$; these two $\triangle$s are in perspective, the line $Oaa'a''$ being their axis of perspective. Hence the line joining their corresponding vertices are concurrent, which proves the Proposition.

Cor.—The two systems of $\triangle$s, viz., that formed by the lines ab, $a'b'$, $a''b''$; bc, $b'c'$, $b''c''$; ca, $c'a'$, $c''a''$; and the system abc, $a'b'c'$, $a''b''c''$, have corresponding properties—namely, the three axes of perspective of either system meet in the centre of perspective of the other system.

Prop. 15.—We shall conclude this section with the solution of a few Problems:—

(1). *To describe a rectangle of given area, whose four sides shall pass through four given points.*

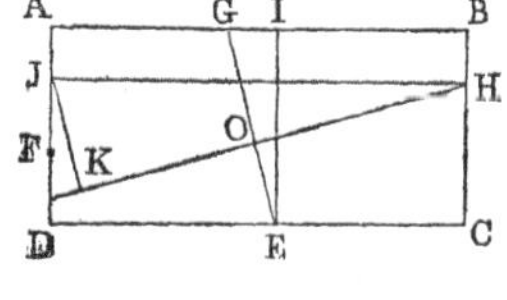

Analysis.—Let ABCD be the required rectangle; E, F, G, H the four given points. Through E draw EI ∥ to AD; and through H draw HJ ∥ to AB, and HO ⊥ to EG; and draw JK ⊥ to HO produced.

Now it is evident that the △s EIG, JHK, are equiangular; ∴ the rectangle EI . JH = EG . HK; but EI . JH = area of rectangle, and is given; ∴ the rectangle EG . HK is given, and EG is given; ∴ HK is given. Hence the line KJ is given in position; and since the angle FJH is right, the semicircle described on HF will pass through J, and is given in position. Hence the point J, being the intersection of a given line and a given ⊙, is given in position; therefore the line FJ is given in position.

(2). *Given the base of a triangle, the perpendicular, and the sum of the sides, to construct it.*

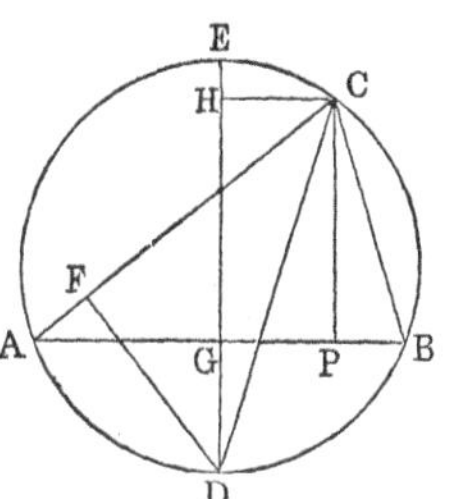

Analysis.—Let ABC be the △, CP the ⊥; and let DE be the diameter of the circumscribed ⊙, which is ⊥ to AB; draw CH ∥ to AB.

Now the rectangle DH . EG is equal to the square of half sum of the sides (IV., 8); ∴ DH.EG is given; and DG . GE = square of GB, and is given. Hence the ratio of DH . GE : DG . GE is given; ∴ the ratio of DH : DG is given. Hence the ratio of GH : DG is given; but GH is = to the ⊥, and is given; hence DG is given; then, if AB be given in position, the point D is given; ∴ the ⊙ ADB is given in position, and CH at a given distance from AB is given in position. Hence the point C is given in position.

The method of construction derived from this analysis is evident.

Cor.—If the base, the perpendicular, and the difference of the sides be given, a slight modification of the foregoing analysis will give the solution.

(3). *Given the base of a triangle, the vertical angle, and the bisector of the vertical angle, to construct the triangle.*

Analysis.—Let ABC be the required △, and let the base AB be given in position; then, since AB is given in position and magnitude, and the ∠ ACB is given in magnitude, the circumscribed ⊙ is given in position. Let CD, the bisector of the vertical ∠, meet the circumscribed ⊙ in E, then E is a given point. Hence EB is given in magnitude.

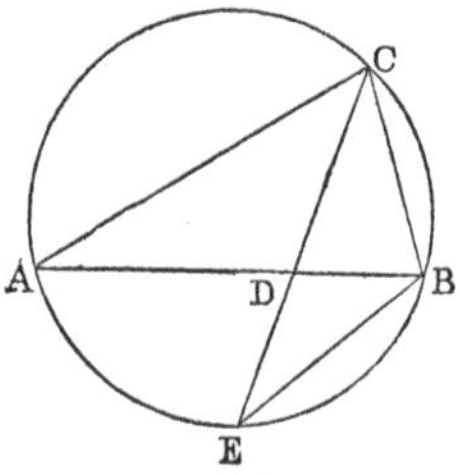

Now ED . EC = EB² (III., 20, *Cor.* 2); ∴ the rectangle ED . EC is given, and CD is given (Hyp.). Hence ED, EC are each given, and the ⊙ described from E as centre, with EC as radius, is given in position. Hence the point C is given, and the method of construction is evident.

Cor.—From the foregoing we may infer the method of solving the Problem: Given the base, vertical angle, and external bisector of the vertical angle.

(4). *Given the base of a triangle, the difference of the base angles, and the difference of the sides, to construct it.*

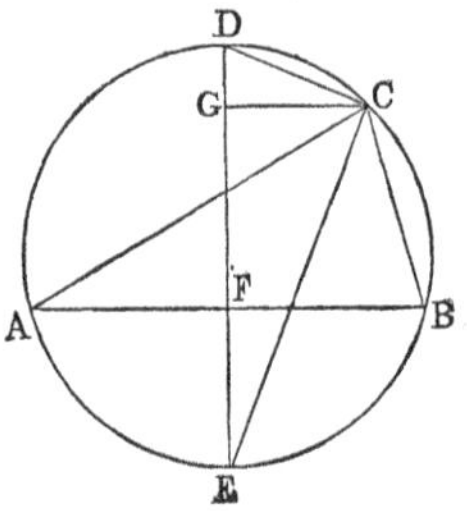

Analysis.—Let ABC be the required △; then the rectangle EF . GD = the square of half the difference of the sides (IV., 8); ∴ EF . GD is given; and EF . FD = FB² is given. Hence the ratio of EF . GD : EF . FD is given. Hence the ratio of FD : GD is given.

Again, the ∠ CED = half the difference of the base ∠ s,

and is given; and DCE is a right ∠; ∴ △ DCE is given in species, and CGD is equiangular to DCE; ∴ CGD is given in species; ∴ the ratios of GD : DC and of DC : DE are given. Hence the ratio of FD : DE is given; therefore the ratio of DF : FE is given, and their rectangle is given. Hence DF and FE are each given. Hence the Proposition is solved.

Cor.—In a like manner we may solve the Problem: Given the base, the difference of the base angles, and the sum of the sides to construct the triangle.

(5). *To construct a quadrilateral of given species whose four sides shall pass through four given points.*

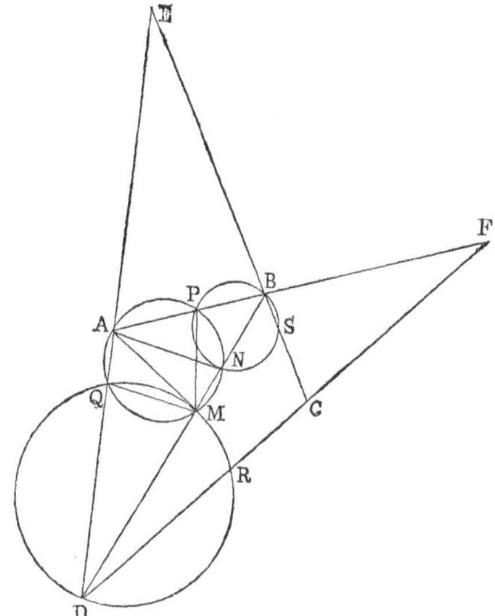

Analysis.—Let ABCD be the required quadrilateral, P, Q, R, S the four given points. Let E, F be the extremities of the third diagonal. Now, let us consider the △ ADF; it is evidently given in species, and PQR is an inscribed triangle given in species. Hence, if M be the point of intersection of circles described about the △s PAQ, QDR, the △ MAD is given in species.—See Demonstration of (III., 17).

In like manner, if N be the point of intersection of the ⊙s about the △s QAP, PBS, the △ ABN is given in species. Hence the ratios AM : AD and AN : AB are given; but the ratio of AB to AD is given, because the figure ABCD is given in species. Hence the ratio of AM : AN is given; and M, N are given points; therefore the locus of A is a circle (7); and where this circle intersects the circle PAQ is a given point. Hence A is given.

Cor.—A suitable modification of the foregoing, and making use of (III., 16), will enable us to solve the cognate Problem—To describe a quadrilateral of given species whose four vertices shall be on four given lines.

(6). *Given the base of a triangle, the difference of the base angles, and the rectangle of the sides, construct it.*

(7). *Given the base of a triangle, the vertical angle, and the ratio of the sum of the sides to the altitude: construct it.*

SECTION II.

Centres of Similitude.

Def.—*If the line joining the centres of two circles be divided internally and externally in the ratio of the radii of the circles, the points of division are called, respectively, the* internal *and the* external *centre of similitude of the two circles.*

From the Definitions it follows that the point of contact of two circles which touch *externally* is an *internal* centre of similitude of the two circles; and the point of contact of two circles, one of which touches another *internally*, is an *external* centre of similitude. Also, since a right line may be regarded as an infinitely large circle, whose centre is at infinity in the direction perpendicular to the line, the centres of similitude of a line and a circle are the two extremities of the diameter of the circle which is perpendicular to the line.

Prop. 1.—*The direct common tangent of two circles passes through their external centre of similitude.*

Dem.—Let O, O′ be the centres of the ⊙s; P, P′ the points of contact of the common tangent; and let PP′ and OO′ produced meet in T; then, by similar △s,

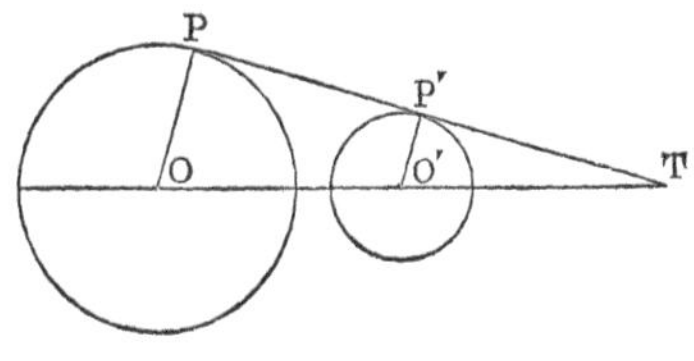

$$OT : O'T :: OP : O'P'.$$

Hence the line OO′ is divided externally in T in the ratio of the radii of the circles; and therefore T is the external centre of similitude.

Cor. 1.—It may be proved, in like manner, that the transverse common tangent passes through the internal centre of similitude.

Cor. 2.—The line joining the extremities of parallel radii of two ⊙s passes through their external centre of similitude, if they are turned in the same direction; and through their internal centre, if they are turned in opposite directions.

Cor. 3.—The two radii of one ⊙ drawn to its points of intersection, with any line passing through either centre of similitude, are respectively ∥ to the two radii of the other ⊙ drawn to its intersections with the same line.

Cor. 4.—All lines passing through a centre of similitude of two ⊙s are cut in the same ratio by the ⊙s.

Prop. 2.—*If through a centre of similitude of two circles we draw a secant cutting one of them in the points* R, R′, *and the other in the corresponding points* S, S′; *then* *the rectangles* OR . OS′, OR′ . OS *are constant and equal.*

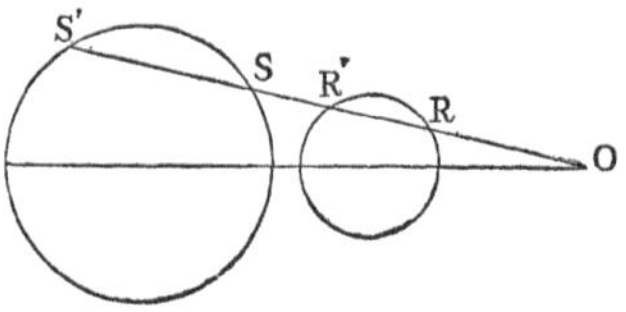

S′ S R′ R O

Dem.—Let a, b denote the radii of the circles; then we have (*Cor.* 3, Prop. 2),

$$a : b :: \text{OS} : \text{OR};$$

therefore $a : b :: \text{OS} . \text{OS}' : \text{OR} . \text{OS}'$;

but OS . OS′ = square of the tangent from O to the circle whose radius is a, and is therefore constant. Hence, since the three first terms of the proportion are constant, the fourth term is constant.

In like manner, it may be proved that OR′ . OS is a fourth proportional to a, b and OS . OS′; ∴ OR′ . OS is constant.

Prop. 3.—*The six centres of similitude of three circles lie three by three on four lines, called axes of similitude of the circles.*

Dem.—Let the radii of the ⊙s be denoted by a, b, c, their centres by A, B, C; the external centres of similitude by A′, B′, C′, and their internal centres by A″, B″, C″. Now, by Definition,

$$\frac{\text{AC}'}{\text{C}'\text{B}} = -\frac{a}{b};$$

$$\frac{\text{BA}'}{\text{A}'\text{C}} = -\frac{b}{c};$$

$$\frac{\text{CB}'}{\text{B}'\text{A}} = -\frac{c}{a}.$$

Hence the product of the three ratios on the right is negative unity; and therefore the points A′, B′, C′ are collinear (*Cor.* 1, Prop. 4).

Again, let us consider the system of points A′, B″, C′. We have, as before,

$$\frac{AC'}{C'B} = -\frac{a}{b};$$

$$\frac{BA''}{A''C} = \frac{b}{c};$$

$$\frac{CB''}{B''A} = \frac{c}{a}.$$

Hence the product of the ratios in this case also is negative unity; and ∴ A″, B″, C′ are collinear; and the same holds for A′, B″, C″; A″, B′, C″. Hence the collinearity of centres of similitude will be one external and two internal, or three external centres of similitude.

Cor. 1.—If a variable ⊙ touch two fixed ⊙s, the line joining the points of contact passes through a fixed point, namely—a centre of similitude of the two ⊙s; for the points of contact are centres of similitude.

Cor. 2.—If a variable ⊙ touch two fixed ⊙s, the tangent drawn to it from the centre of similitude through which the chord of contact passes is constant.

Prop. 4.—*If two circles touch two others, the radical axis of either pair passes through a centre of similitude of the other pair.*

Dem.—Let the two ⊙s X, Y touch the two ⊙s W, V; let R, R′ be their points of contact with W, and S, S′ with V. Now, consider the three ⊙s X, W, Y; R, R′ are internal centres of similitude. Hence the line RR′ passes through the external centre of similitude of X and Y.

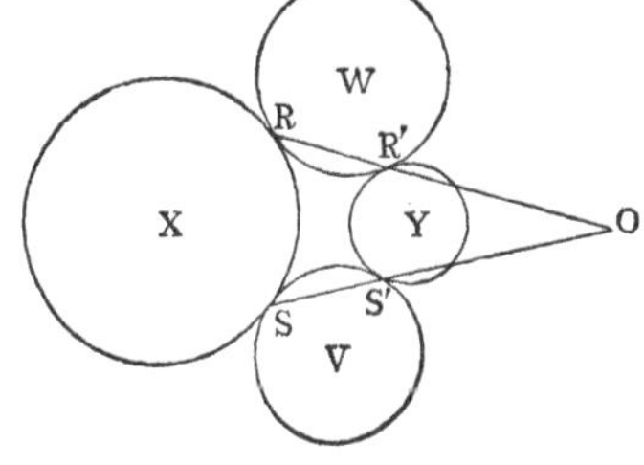

In like manner, the line SS′ passes through the same centre of similitude. Hence the point O, where these lines meet, will be the external centre of similitude of X and Y; and ∴ the rectangle OR . OR′ = OS . OS′ (Prop. 2); ∴ tangent from O to W = tangent from O to V, hence the radical axis of W and V passes through O.

Def.—*The circle on the interval, between the centres of similitude of two circles as diameter, is called their circle of similitude.*

Prop. 5.—*The circle of similitude of two circles is the locus of the vertex of a triangle whose base is the interval between the centres of the circles, and the ratio of the sides that of their radii.*

Dem.—When the base and the ratio of the sides are given, the locus of the vertex (see Prop. 7, Section I) is the ⊙ whose diameter is the interval between the points in which the base is divided in the given ratio internally and externally; that is, in the present case, the ⊙ of similitude.

Cor. 1.—If from any point in the ⊙ of similitude of two given ⊙s lines be drawn to their centres, these lines are proportional to the radii of the two given ⊙s.

Cor. 2.—If, from any point in the ⊙ of similitude of two given ⊙s, pairs of tangents be drawn to both ⊙s, the angle between one pair is equal to the angle between the other pair.

This follows at once from *Cor.* 1.

Cor. 3.—The three ⊙s of similitude of three given ⊙s taken in pairs are coaxal.

For, let P, P′ be the points of intersection of two of the ⊙s of similitude, then it is evident that the lines drawn from either of these points to the centres of the three given ⊙s are proportional to the radii of the given ⊙s. Hence the third ⊙ of similitude must pass through the points P, P′. Hence the ⊙s are coaxal.

Cor. 4.—The centres of the three ⊙s of similitude of three given ⊙s taken in pairs are collinear.

SECTION III.

Theory of Harmonic Section.

Def. — *If a line* AB *be divided internally in the point* C, *and externally in the point* D, *so that the ratio* AC : CB = – ratio AD : DB; the points C and D are called harmonic conjugates to the points A, B.

Since the segments AC, CB are measured in the same direction, the ratio AC : CB is positive; and AD, DB being measured in opposite directions, their ratio is negative. This explains why we say AC : CB = – AD : DB. We shall, however, usually omit the sign minus, unless when there is special reason for retaining it.

Cor.—The centres of similitude of two given circles are harmonic conjugates, with respect to their centres.

Prop. 1.—*If* C *and* D *be harmonic conjugates to* A *and* B, *and if* AB *be bisected in* O, *then* OB *is a geometric mean between* OC *and* OD.

Dem.— AC : CB : : AD : DB;

$$\therefore \quad \frac{AC - CB}{2} : \frac{AC + CB}{2} :: \frac{AD - DB}{2} : \frac{AD + DB}{2};$$

or OC : OB : : OB : OD.

Hence OB is a geometric mean between OC and OD.

Prop. 2.—*If* C *and* D *be harmonic conjugates to* A *and* B, *the circles described on* AB *and* CD *as diameters intersect each other orthogonally.*

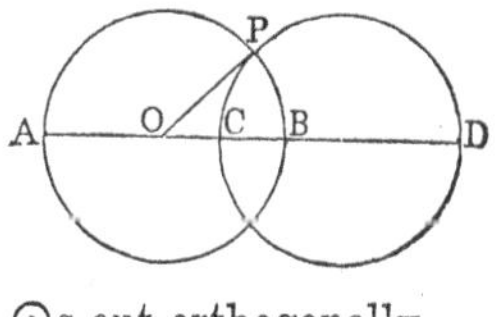

Dem.—Let the ⊙s intersect in P, bisect AB in O; join OP; then, by Prop. 2, we have OC . OD = OB^2 = OP^2. Hence OP is a tangent to the circle CPD, and therefore the ⊙s cut orthogonally.

Cor. 1.—Any ⊙ passing through the points C and D will be cut orthogonally by the ⊙ described on AB as diameter.

Cor. 2.—The points C and D are inverse points with respect to the ⊙ described on AB as diameter.

Def.—*If* C *and* D *be harmonic conjugates to* A *and* B, AB *is called a* harmonic mean *between* AC *and* AD.

Observation.—This coincides with the the algebraic Definition of *harmonic mean.*

For AC, AB, AD being three magnitudes, we have

$$AC : CB :: AD : BD;$$

therefore $$AC : AD :: CB : BD;$$

that is, the 1st is to the 3rd as the difference between the 1st and 2nd is to the difference between the 2nd and 3rd, which is the algebraic Definition.

Cor.—In the same way it can be seen that DC is a harmonic mean between DA and DB.

Prop. 3.—*The Arithmetic mean is to the Geometric mean as the Geometric mean is to the Harmonic mean.*

Dem.—Upon AB as diameter describe a ⊙; erect EF at right angles to AB through C; draw tangents to the ⊙ at E, F, meeting in D; then, since the △ OED is right-angled at E, and EC is ⊥ to OD, we have OC . OD = OE^2 = OB^2. Hence, by Prop. 1, C and D are harmonic conjugates to A and B. Again, from the same △, we have OD : DE : : DE : DC; but OD = $\frac{1}{2}$(DA + DB) = arithmetic mean between DA and DB; and DE is the geometric mean and DC the harmonic mean between DA and DB.

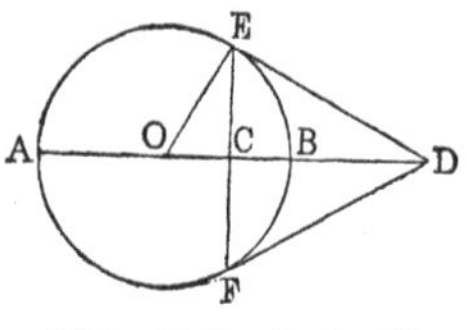

Cor.—The reciprocals of the three magnitudes DA, DO, DB are respectively DB, DC, DA, with respect to DE^2; but DA, DO, DB are in arithmetical progression.

Hence the reciprocals of lines in arithmetical progression are in harmonical progression.

Prop. 4.—*Any line cutting a circle, and passing through a fixed point, is cut harmonically by the circle, the point, and the polar of the point.*

Let D be the point, EF its polar, DGH a line cutting the ⊙ in the points G and H, and the polar of D in the point J; then the points J, D will be harmonic conjugates to H and G.

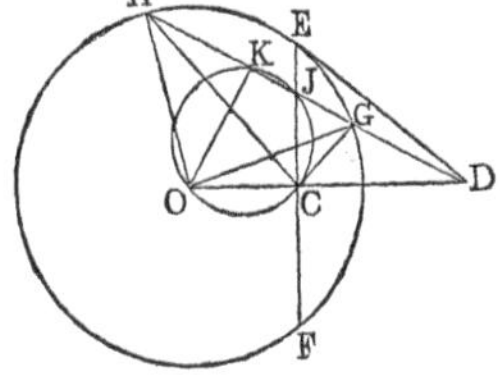

Dem.—Let O be the centre of the ⊙; from O let fall the ⊥ OK on HD; then, since K and C are right ∠s, OKJC is a quadrilateral in a ⊙; ∴ OD.DC=KD.DJ; but OD.DC=DE^2; ∴ KD.DJ = DE^2. Hence KD : DE :: DE : DJ; and since KD, DE are respectively the arithmetic mean and the geometric mean between DG and DH, DJ (Prop. 3.) will be the harmonic mean between DG and DH.

The following is the proof usually given of this Proposition:—Join OH, OG, CH, CG. Now OD . DC = DE^2 = DH . DG; ∴ the quadrilateral HOCG is inscribed in a ⊙; ∴ the angle OCH = OGH; and DCG = OHD; but OGH = OHD; ∴ OCH = DCG. Hence HCJ = GCJ; hence CJ and CD are the internal and external bisectors of the vertical angle GCH of the triangle GCH; therefore the points J and D are harmonic conjugates to the points H and G. Q. E. D.

Cor. 1.—If through a fixed point D any line be drawn cutting the ⊙ in the points G and H, and if DJ be a harmonic mean between DG and DH, the locus of J is the polar of D.

Cor. 2.—In the same case, if DK be the arithmetic mean between DG and DH, the locus of K is a ⊙, namely, the ⊙ described on OD as diameter, for the ∠ OKD is right.

Prop. 5.—*If* ABC *be a triangle,* CE *a line through the vertex parallel to the base* AB; *then any transversal through* D, *the middle of* AB, *will meet* CE *in a point, which will be the harmonic conjugate of* D, *with respect to the points in which it meets the sides of the triangle.*

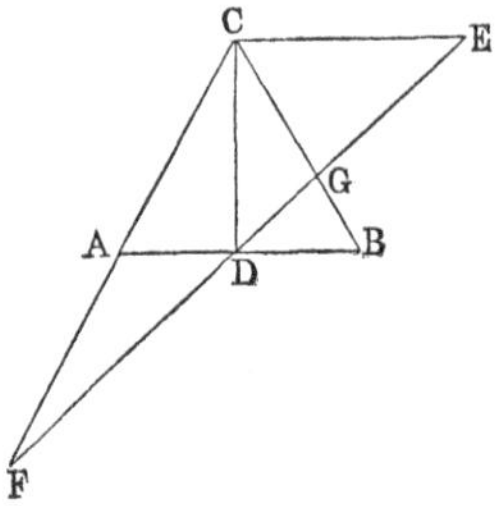

Dem.—From the similar △s FCE, FAD we have EF : FD : : CE : AD; but AD = DB; ∴ EF : FD :: CE : DB.

Again, from the similar △s CEG, BDG, we have CE : DB : : EG : GD;

therefore EF : FD : : EG : GD. Q.E.D.

Defs.—*If we join the points* C, D (see last diagram), *the system of four lines* CA, CD, CB, CE *is called a harmonic pencil; each of the four lines is called a ray; the point* C *is called the vertex of the pencil; the alternate rays* CD, CE *are said to be harmonic conjugates with respect to the rays* CA, CB. *We shall denote such a pencil by the notation* (C . FDGE), *where C is the vertex;* CF, CD, CG, CE *the rays.*

Prop. 6.—*If a line* AB *be cut harmonically in* C *and* D, *and a harmonic pencil* (O . ABCD) *formed by joining the points* A, B, C, D *to any point* O; *then, if through* C, *a parallel to* OD, *the ray conjugate to* OC *be drawn, meeting* OA, OB *in* G *and* H, GH *will be bisected in* C.

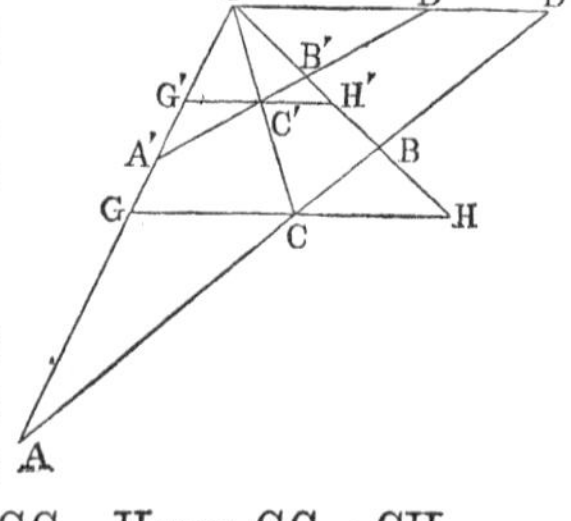

Dem.—

OD : CH : : DB : BC;

and OD : GC : : DA : AC;

but DB : BC : : DA : AC;

∴ OD : CH : : OD : GC. Hence GC = CH.

Cor.—Any transversal A′B′C′D′ cutting a harmonic pencil is cut harmonically.

For, through C′ draw G′H′ ∥ to GH; then, by Prop. 3, Section I., G′C′ : C′H′ : : GC : CH; ∴ G′C′ = C′H′. Hence A′B′C′D′ is cut harmonically.

Prop. 7.—*The line joining the intersection of two opposite sides of a quadrilateral with the intersection of its diagonals forms, with the third diagonal, a pair of rays, which are harmonic conjugates with these sides.*

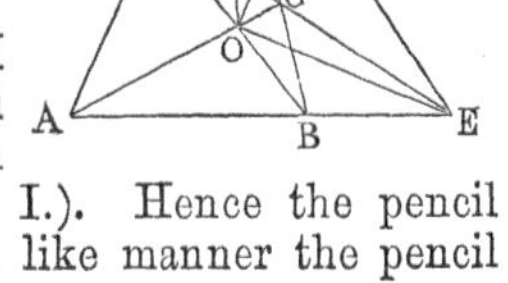

Let ABCD be the quadrilateral whose two sides AD, BC meet in F; then the line FO, and the third diagonal FE, form a pair of conjugate rays with FA and FB.

Dem.—Through O draw OH ∥ to AD; meet BC in G, and the third diagonal in H. Then OG = GH (Prop. 8, Section I.). Hence the pencil (F . AOBE) is harmonic. In like manner the pencil (E . AODF) is harmonic.

Prop. 8. — *If four collinear points form a harmonic system, their four polars with respect to any circle form a harmonic pencil.*

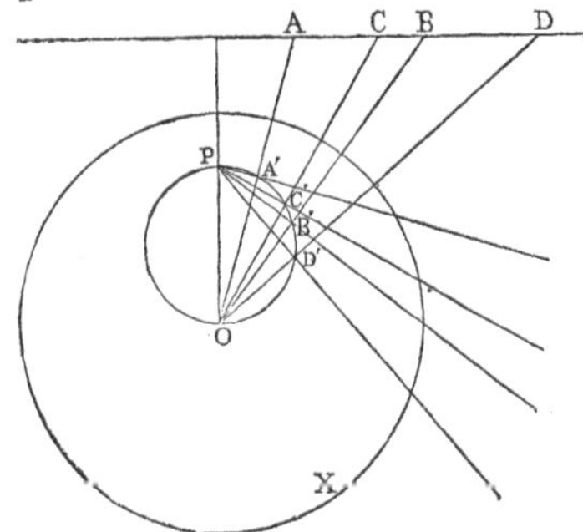

Let A, C, B, D be the four points, P the pole of their line of collinearity with respect to the ⊙ X; let

O be the centre of X. Join OA, OB, OC, OD, and let fall the ⊥s PA′, PB′, PC′, PD′ on these lines; then, by Prop. 25, Section I., Book III., PA′, PB′, PC′, PD′ are the polars of the points A, B, C, D; and since the angles at A′, C′, B′, D′ are right, the ⊙ described on OP as diameter will pass through these points; and since the system A, B, C, D is harmonic, the pencil (O . ABCD) is harmonic; but the angles between the rays OA, OB, OC, OD are respectively equal to the angles between the rays PA′, PB′, PC′, PD′ (III., xxi.). Hence the pencil (P . A′B′C′D′) is harmonic.

Def.—*Four points in a circle which connect with any fifth point in the circumference by four lines, forming a harmonic pencil, are called a harmonic system of points on the circle.*

Prop. 9.—*If from any point two tangents be drawn to a circle, the points of contact and the points of intersection of any secant from the same point form a harmonic system of points.*

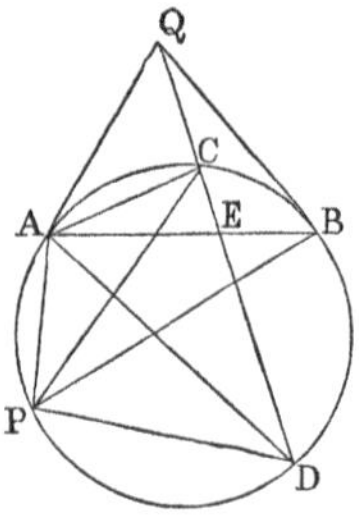

Dem.—Let Q be the point, QA, QB tangents, QCD the secant; take any point P in the circumference of the ⊙, and join PA, PC, PB, PD; then, since AB is the polar of Q, the points E, Q are harmonic conjugates to C and D; ∴ the pencil (A . QCED) is harmonic; but the pencil (P . ACBD) is equal to the pencil (A . QCED), for the angles between the rays of one equal the angles between the rays of the other; therefore the pencil (P . ACBD) is harmonic. Hence A, C, B, D form a harmonic system of points.

Cor. 1.—If four points on a ⊙ form a harmonic system, the line joining either pair of conjugates passes through the pole of the line joining the other pair.

Cor. 2.—If the angular points of a quadrilateral inscribed in a ⊙ form a harmonic system, the rectangle contained by one pair of opposite sides is equal to the rectangle contained by the other pair.

Prop. 10.—*If through any point* O *two lines be drawn cutting a circle in four points, then joining these points both directly and transversely; and if the direct lines meet in* P *and the transverse lines meet in* Q, *the line* PQ *will be the polar of the point* O.

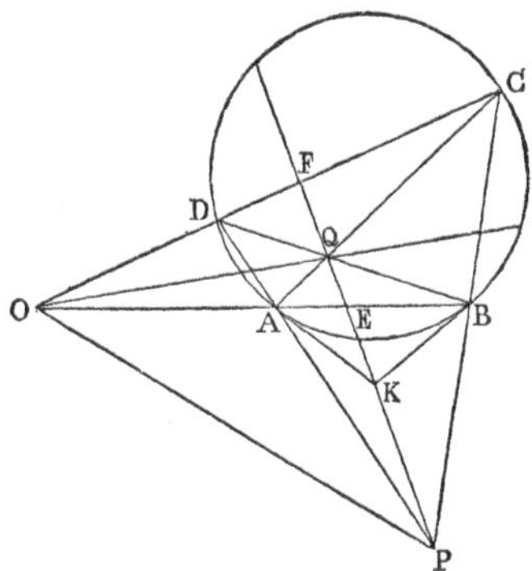

Dem.—Join OP; then the pencil (P . OAEB) is harmonic (Prop. 7); ∴ the points O, E are harmonic conjugates to the points A, B. Hence the polar of O passes through E (Prop. 4). In like manner, the polar of O passes through F; ∴ the line PQ, which passes through the points E and F, is the polar of O. Q. E. D.

Cor. 1.—If we join the points O and Q, it may be proved in like manner that OQ is the polar of P.

Cor. 2.—Since PQ is the polar of O, and OQ the polar of P, then (*Cor.* 1, Prop. 16, Section I., Book III.) OP is the polar of Q.

Def.—*Triangles such as* OPQ, *which possess the property that each side is the polar of the opposite angular point with respect to a given circle, are called self-conjugate triangles with respect to the circle. Again, if we consider the four points* A, B, C, D, *they are joined by three pairs of lines, which intersect in the three points* O, P, Q *respectively; then, on account of the harmonic properties of the quadrilateral* ABCD *and the triangle* OPQ, *I propose to call* OPQ *the harmonic triangle of the quadrilateral.*

Prop. 11.—*If a quadrilateral be inscribed in a circle, and at its angular points four tangents be drawn, the six points of intersection of these four tangents lie in pairs on the sides of the harmonic triangle of the inscribed quadrilateral.*

Dem.—Let the tangents at A and B meet in K (see fig., last Prop.); then the polar of the point K passes through O. Hence the polar of O passes through K; therefore the point K lies on PQ. In like manner, the tangents at C and D meet on PQ. Hence the Proposition is proved.

Cor. 1.—Let the tangents at B and C meet in L, at C and D in M, at A and D in N; then the quadrilateral KLMN will have the lines KM (PQ) and LN (OQ) as diagonals; therefore the point Q is the intersection of its diagonals. Hence we have the following theorem:—*If a quadrilateral be inscribed in a circle, and tangents be drawn at its angular points, forming a circumscribed quadrilateral, the diagonals of the two quadrilaterals are concurrent, and form a harmonic pencil.*

Cor. 2.—The tangents at the points B and D meet on OP, and so do the tangents at the points A and C. Hence the line OP is the third diagonal of the quadrilateral KLMN; and the extremities of the third diagonal are the poles of the lines BD, AC. Now, since the lines BD, AC are harmonic conjugates to the lines QP, QO, the poles of these four lines form a harmonic system of points. Hence we have the following theorem:—*If tangents be drawn at the angular points of an inscribed quadrilateral, forming a circumscribed quadrilateral, the third diagonals of these two quadrilaterals are coincident, and the extremities of one are harmonic conjugates to the extremities of the other.*

SECTION IV.

Theory of Inversion.*

Def.—*If* X *be a circle,* O *its centre,* P *and* Q *two points on any radius, such that the rectangle* OP . OQ = *square of the radius, then* P *and* Q *are called* inverse *points with respect to the circle.*

If one of the points, say Q, describe any curve, a circle for instance, the other point P will describe the inverse curve.

We have already given in Book III., Section I., Prop. 20, the inversion of a right line; in Book IV., Section I., Prop. 7, one of its most important applications. This section will give a systematic account of this method of transformation, one of the most elegant in Geometry.

Prop. 1.—*The inverse of a circle is either a line or a circle, according as the centre of inversion is on the circumference of the circle or not on the circumference.*

Dem.—We have proved the first case in Book III.; the second is proved as follows:—Let Y be the ⊙ to be inverted, O the centre of inversion; take any point P in Y; join OP, and make OP . OQ = constant (square of radius of inversion); then Q is the inverse of P: it is required to find the locus of Q. Let OP produced, if necessary, meet the ⊙ Y again

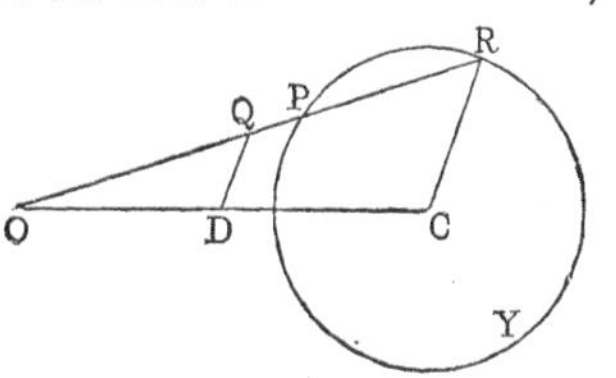

* This method, one of the most important in the whole range of Geometry, is the joint discovery of Doctors Stubbs and Ingram, Fellows of Trinity College, Dublin (see the *Transactions* of the Dublin Philosophical Society, 1842). The next writer that employed it is Sir William Thomson, who by its aid gave geometrical proofs of some of the most difficult propositions in the Mathematical Theory of Electricity (see Clerk Maxwell on "Electricity," Vol. I., Chapter XI.).

at R; then the rectangle OP . OR = square of tangent from O (III. xxxvi.), and ∴ = constant, and OP . OQ is constant (hyp.); ∴ the ratio of OP . OR : OP . OQ is constant: hence the ratio of OR : OQ is constant. Let C be the centre of Y; join OC, CR, and draw QD ∥ to CR. Now OR : OQ :: CR : QD; ∴ the ratio of CR : QD is constant, and CR is constant; ∴ QD is constant. In like manner OD is constant; ∴ D is a given point; ∴ the locus of Q is a ⊙, whose centre is the given point D, and whose radius is DQ.

Cor. 1.—The centre of inversion O is the centre of similitude of the original circle Y, and its inverse.

Cor. 2.—The circle Y, its inverse, and the circle of inversion are coaxal. For if the ⊙ Y be cut in any point by the ⊙ of inversion, the ⊙ inverse to Y will pass through that point.

Prop. 2.—*If two circles, or a line and a circle, touch each other, their inverses will also touch each other.*

Dem.—If two ⊙s, or a line and a ⊙ touch each other, they have two consecutive points common; hence their inverses will have two consecutive points common, and therefore they touch each other.

Prop. 3.—*If two circles, or a line and a circle, intersect each other, their angle of intersection is equal to the angle of intersection of their inverses.*

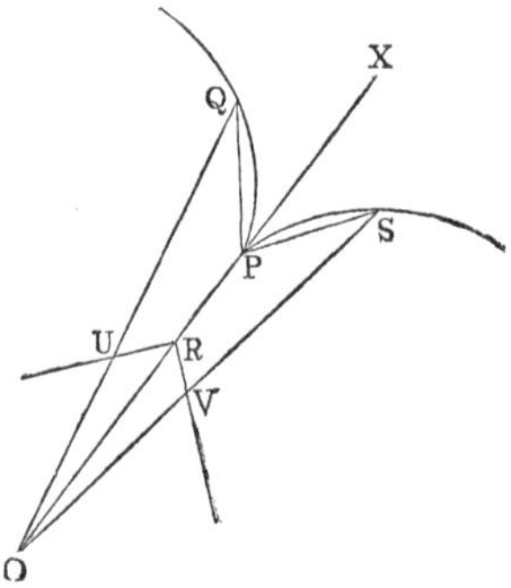

Dem.—Let PQ, PS be parts of two ⊙s intersecting in P; let O be the centre of inversion. Join OP; let Q and S be two points on the ⊙s very near P. Join OQ, OS, PQ, PS; and let R, U, V be the inverses of the points P, Q, S. Join UR, VR, and produce OP to X. Now, from the construction, U and V are points on the inverses of the ⊙s PQ, PS. And since the rectangle OP . OR = rectangle OQ . OU, the quadrilateral

RPQU is inscribed in a ⊙; ∴ the ∠ ORU = OQP; and when Q is infinitely near P, the ∠ OQP = QPX; ∴ the ∠ ORU is ultimately = QPX. In like manner, the ∠ ORV is ultimately equal to the ∠ SPX; ∴ the ∠ URV is ultimately equal to the ∠ QPS. Now QP, SP are ultimately tangents to their respective circles, and ∴ the ∠ QPS is their angle of intersection, and URV is the angle of intersection of the inverses of the circles. Hence the Proposition is proved.

Prop. 4.—*Any two circles can be inverted into themselves.*

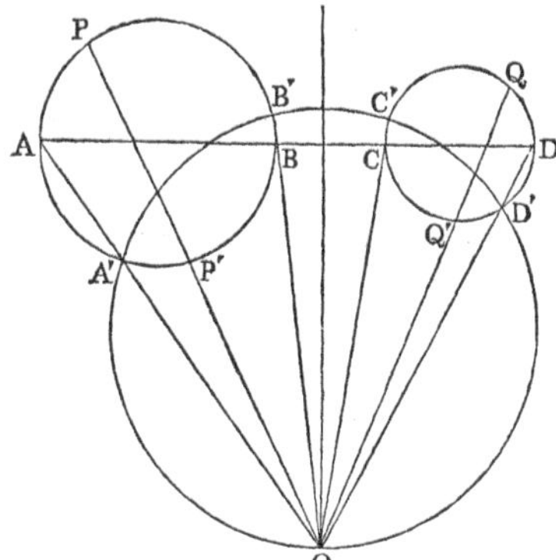

Dem.—Take any point O in the radical axis of the two ⊙s; and from O draw two lines OPP′, OQQ′, cutting the ⊙s in the points P, P′, Q, Q′; then the rectangle OP . OP′ = the rectangle OQ . OQ′ = square of tangent from O to either of the circles, and ∴ equal to the square of the radius of the circle whose centre is O, and which cuts both circles orthogonally. Hence the points P′, Q′ are the inverses of the points P and Q with respect to the orthogonal circle; and therefore while the points P, Q move along their respective circles, their inverses, the points P′, Q′, move along other parts of the same circles.

Cor. 1.—The circle of self-inversion of a given circle cuts it orthogonally.

Cor. 2.—Any three circles can be inverted into themselves, their circle of self-inversion being the circle which cuts the three circles orthogonally.

Cor. 3.—If two circles be inverted into themselves, the line joining their centres, namely ABCD, will be inverted into a circle cutting both orthogonally; for the line ABCD cuts the two circles orthogonally.

Cor. 4.—Any circle cutting two circles orthogonally may be regarded as the inverse of the line passing through their centres.

Cor. 5.—If ABCD be the line passing through the centres of two circles, and A′B′C′D′ any circle cutting them orthogonally; then the points A′, B′, C′, D′ being respectively the inverses of the points A, B, C, D, the four lines AA′, BB′, CC′, DD′ will be concurrent.

Cor. 6.—Any three circles can be inverted into three circles whose centres are collinear.

Prop. 5.—*Any two circles can be inverted into two equal circles.*

Dem.—Let X, Y be the original ⊙s, r and r' their radii; let V, W be the inverse ⊙s, ρ and ρ' their radii; and let O be the centre of inversion, and T, T′ the tangents from O to X and Y, and R the radius of the circle of inversion. Then, from the Demonstration of Prop. 1, we have

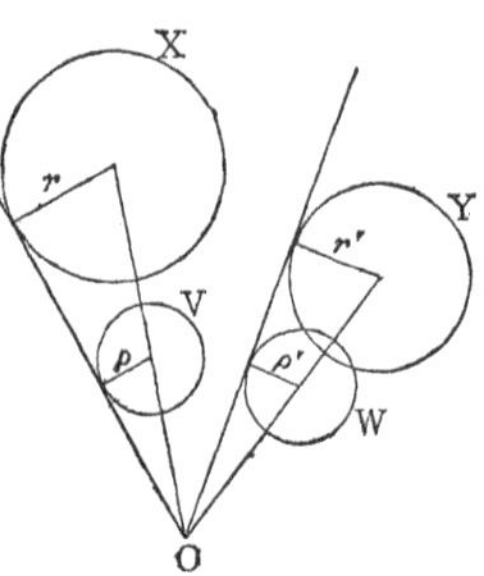

$$r : \rho :: T^2 : R^2;$$

$$r' : \rho' :: T'^2 : R^2.$$

Hence, since $\rho = \rho'$, we have

$$r : r' :: T^2 : T'^2;$$

∴ the ratio of $T^2 : T'^2$ is given; and, consequently, the ratio of $T : T'$ is given. Hence if a point be found, such that the tangents drawn from it to the two ⊙s X, Y will be in the ratio of the square roots of their radii, and if X, Y, be inverted from that point, their inverses will be equal. It will be seen, in the next Section, that the locus of O is a circle coaxal with X and Y.

Cor. 1.—Any three circles can be inverted into three equal circles.

Cor. 2.—Hence can be inferred a method of describing a circle to touch any three circles.

Cor. 3.—If any two circles be the inverses of two others, then any circle touching three out of the four circles will also touch the fourth.

Cor. 4.—If any two points be the inverses of two other points, the four points are concyclic.

Prop. 6.—*If* A *and* B *be any two points,* O *a centre of inversion; and if the inverses of* A, B *be the points* A′, B′, *and* p, p', *the perpendiculars from* O *on the lines* AB, A′B′; *then* AB : A′B′ :: p : p'.

Dem.—Since O is the centre of inversion, we have

$$OA \,.\, OA' = OB \,.\, OB';$$

therefore $$OA : OB :: OB' : OA'.$$

And the angle O is common to the two △s AOB, A′OB′; ∴ the △s are equiangular. Hence the Proposition is proved.

Prop. 7.—*If* A, B, C . . . L *be any number of collinear points, we have*

$$AB + BC + CD \,..\, + LA = 0.$$

(Since LA is measured backwards, it is regarded as negative.) Now, let p be the ⊥ from any point O on the line AL; and, dividing by p, we have

$$\frac{AB}{p} + \frac{BC}{p} + \frac{CD}{p} \,..+ \frac{LA}{p} = 0.$$

Let the whole be inverted from O; and, denoting the

inverses of the points A, B, C . . . L by A′, B′, C′ . . . L′, we have from the last Article the following general theorem:—*If a polygon* A′B′C′ . . . L′ *of any number of sides be inscribed in a circle, and if from any point in its circumference perpendiculars be let fall on the sides of the polygon; then the sum of the quotients obtained by dividing the length of each side by its perpendicular is zero.*

Cor. 1.—Since one of the ⊥s must fall externally on its side of the polygon, while the other ⊥s fall internally, this ⊥ must have a contrary sign to the remainder. Hence the Proposition may be stated thus:—*The length of the side on which the perpendicular falls externally, divided by its perpendicular, is equal to the sum of the quotients arising by dividing each of the remaining sides by its perpendicular.*

Cor. 2.—Let there be only three sides, and let the ⊥s be α, β, γ; then, if a, b, c denote the lengths of the sides, &c.,

$$\frac{a}{\alpha} + \frac{b}{\beta} + \frac{c}{\gamma} = 0.$$

Prop. 8.—*If* A, B, C, D *be four collinear points,* A′, B′, C′, D′ *the four points inverse to them; then*

$$\frac{AC \cdot BD}{AB \cdot CD} = \frac{A'C' \cdot B'D'}{A'B' \cdot C'D'}.$$

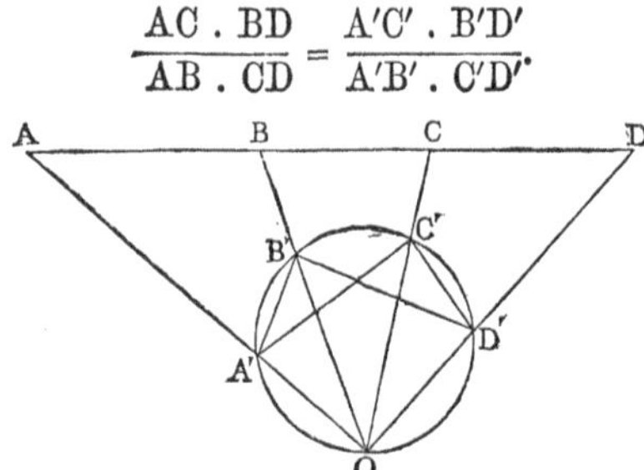

Dem.—Let O be the centre of inversion, and p the ⊥ from O on the line ABCD; and let the ⊥s from O on the lines A′B′, A′C′, B′D′, C′D′ be denoted by α, β,

γ, δ. Then, by Prop. 6, we have the following equalities:—

$$AC = \frac{A'C' \cdot p}{\beta};$$

$$BD = \frac{B'D' \cdot p}{\gamma};$$

$$AB = \frac{A'B' \cdot p}{\alpha};$$

$$CD = \frac{C'D' \cdot p}{\delta}.$$

Hence multiplying, and remembering that the rectangle $\beta\gamma$ is equal to the rectangle $\alpha\delta$ (see Prop. 11, Section I., Book III.), we get

$$\frac{AC \cdot BD}{AB \cdot CD} = \frac{A'C' \cdot B'D'}{A'B' \cdot C'D'}.$$

Cor. 1.—

$$AC \cdot BD : AB \cdot CD : AD \cdot BC$$

$$:: A'C' \cdot B'D' : A'B' \cdot C'D' : A'D' \cdot B'C'.$$

Cor. 2.—If the points A, B, C, D form a harmonic system, the points A′, B′, C′, D′ form a harmonic system. In other words, the inverse of a harmonic system of points forms a harmonic system.

Cor. 3.—If AB = BC; then the points A′, B′, C′, O form a harmonic system of points.

Prop. 9.—*If two circles be inverted into two others, the square of the common tangent of the first pair, divided by the rectangle contained by their diameters, is equal to the square of the common tangent of the second pair, divided by the rectangle contained by their diameters.*

Dem.—Let X, Y be the original ⊙s, X′, Y′ their inverse ⊙s, ABCD the line through the centres of X and Y, and let the inverse of the line ABCD be the ⊙ A′B′C′D′; then, since the line ABCD cuts orthogonally the ⊙s X, Y, its inverse, the ⊙ A′B′C′D′, cuts orthogonally the ⊙s X′, Y′. Let *abcd* be the line through the

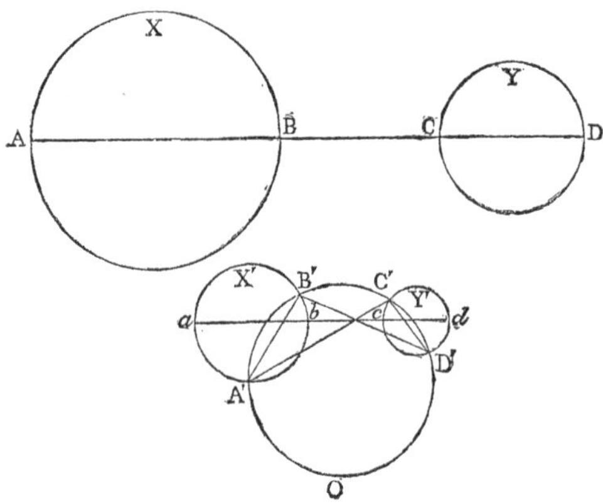

centres of the ⊙s X′, Y′; then *abcd* cuts the ⊙s X′, Y′ orthogonally; hence the ⊙ A′B′C′D′ is the inverse of the line *abcd* with respect to a ⊙ of inversion, which inverts the ⊙s X′, Y′ into themselves (see Prop. 4, *Cor.* 3). Hence, by Prop. 8, each of the ratios

$$\frac{AC \cdot BD}{AB \cdot CD}, \quad \frac{ac \cdot bd}{ab \cdot cd}$$

is equal to the ratio

$$\frac{A'C' \cdot B'D'}{A'B' \cdot C'D'};$$

therefore $$\frac{AC \cdot BD}{AB \cdot CD} = \frac{ac \cdot bd}{ab \cdot cd}.$$

The numerators of these fractions are equal respectively

to the squares of the common tangents of the pairs of circles X, Y; X′, Y′ (see Prop. 8, Section I., Book III). Hence the Proposition is proved.

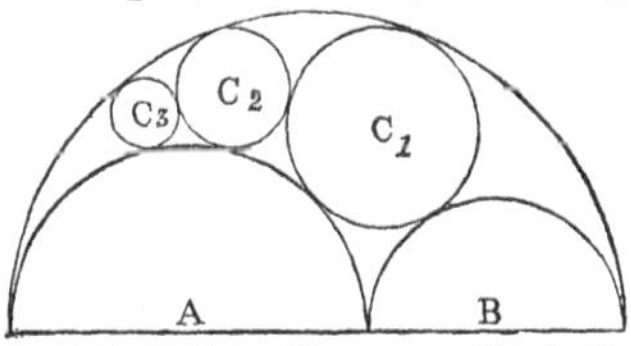

Cor. 1.—If C_1, C_2, C_3, &c., be a series of circles, touching two parallel lines, and also touching each other; then it is evident, by making the diagram, that the square of the direct common tangent of any two of these circles, such as C_m, C_{m+n}, which are separated by $(n - 1)$ circles, is $= n^2$ times the rectangle contained by their diameters. Hence, by inversion and by the theorem of this Article, we have the following theorem:—*If* A *and* B *be any two semicircles in contact with each other, and also in contact with another semicircle, on whose diameter they are described; and if circles* C_1, C_2, C_3 *be described, touching them as in the diagram, the* ⊥ *from the centre of* C_n *on the line* AB = n *times the diameter of* C_n, *where* n *denotes any of the natural numbers* 1, 2, 3, *&c.*

This theorem will immediately follow by completing the semicircles, and describing another system of circles on the other side equal to the system C_1, C_2, C_3, &c., and similarly placed.*

Prop. 10.—*If four circles be all touched by the same circle; then, denoting by* $\overline{12}$, *the common tangent of the* 1*st and* 2*nd, &c.*,

$$\overline{12} \cdot \overline{34} + \overline{14} \cdot \overline{23} = \overline{13} \cdot \overline{24}.$$

Dem.—Let A, B, C, D be four points taken in order on a right line; then, by Prop. 7, Section I., Book II., we have

$$AB \cdot CD + BC \cdot AD = AC \cdot BD.$$

Now, let four arbitrary circles touch the line at the

* The theorem of this Cor. is due to Pappus. See Steiner's *Gesammelte Werke*, Band I., Seite 47.

points A, B, C, D, and let their diameters be δ, δ', δ'', δ'''; then we have

$$\frac{AB \cdot CD}{\sqrt{\delta\delta'} \cdot \sqrt{\delta''\delta'''}} + \frac{BC \cdot AD}{\sqrt{\delta'\delta''} \cdot \sqrt{\delta\delta'''}} = \frac{AC \cdot BD}{\sqrt{\delta\delta''} \cdot \sqrt{\delta'\delta'''}};$$

and by the last Proposition each of the fractions of this equation remains unaltered by inversion. Hence, if the diameters of the inverse circles be denoted by d, d', d'', d''', and their common tangents by $\overline{12}$, &c., we get

$$\frac{\overline{12} \cdot \overline{34}}{\sqrt{dd'} \cdot \sqrt{d''d'''}} + \frac{\overline{23} \cdot \overline{41}}{\sqrt{d'd''} \cdot \sqrt{d'''d}} = \frac{\overline{13} \cdot \overline{24}}{\sqrt{dd''} \cdot \sqrt{d'd'''}}.$$

Hence $$\overline{12} \cdot \overline{34} + \overline{23} \cdot \overline{14} = \overline{13} \cdot \overline{24}.\text{*}$$

Cor. 1.—If four arbitrary circles touch a given circle at a harmonic system of points; then

$$\overline{12} \cdot \overline{34} = \overline{23} \cdot \overline{14}.$$

Cor. 2.—The theorem of this Proposition may be written in the form

$$\overline{12} \cdot \overline{34} + \overline{23} \cdot \overline{14} + \overline{31} \cdot \overline{24} = 0;$$

and in this form it proves at once the property of the "Nine-points Circle." For, taking the ⊙s 1, 2, 3, 4 to be the inscribed and escribed ⊙s of the △, and remembering that when ⊙s touch a line on different sides, we are, in the application of the foregoing theorem, to use transverse common tangents. Hence, making use of the results of Prop. 1, Section I., Book IV., we get

$$\overline{12} \cdot \overline{34} + \overline{23} \cdot \overline{14} + \overline{31} \cdot \overline{24}$$
$$= b^2 - c^2 + c^2 - a^2 + a^2 - b^2 = 0.$$

Hence the ⊙s 1, 2, 3, 4, are all touched by a fifth ⊙.

This theorem is due to Feuerbach. The following simple proof of this now celebrated theorem was pub-

* This extension of Ptolemy's Theorem first appeared in a Paper of mine in the *Proceedings* of the Royal Irish Academy, 1866.

lished by me in the *Quarterly Journal* for February, 1861 :—

"*If* ABC *be a plane triangle, the circle passing through the feet of its perpendiculars touches its inscribed and escribed circles.*"

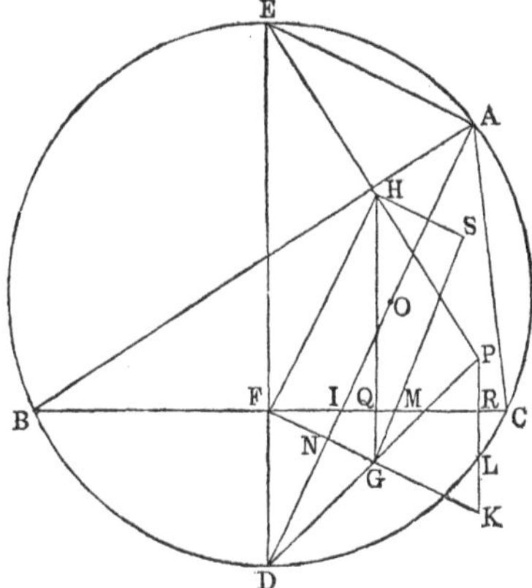

Dem.—Let the inscribed and escribed ⊙s be denoted by O, O′, O″, O‴, the circumscribed ⊙ by X, and the ⊙ through feet of ⊥s by Σ. Now, if P be the intersection of ⊥s, and if the lower segments of ⊥s be produced to meet X, the portions intercepted between P and X are bisected by the sides of ABC (Prop. 13, Section I., Book III.). Hence Σ passes through the points of bisection, and therefore P is the external centre of similitude of X and Σ.

Let DE be the diameter of X, which bisects BC. Join PD, PE, and bisect them in G and H; then Σ must pass through the points G and H; and since GH is ∥ to DE, GH must be the diameter of Σ; and since Σ passes through F, the middle point of BC (see Prop. 5, Section I., Book IV.), the ∠ GFH is right. Again, if from the point D three ⊥s be let fall on the sides of ABC, their feet are collinear, and the line of collinearity evidently is ⊥ to AD and it bisects PD (see Prop. 14, Section I., Book III.). Hence FG is the line of collinearity, and FG is ⊥ to AD. Let M be the point of contact of O with BC; join GM, and let fall the ⊥ HS. Now, since FM is a tangent to O, if from N we draw another tangent to O, we have $FM^2 = FN^2$ + square of tangent from N (Prop. 21, Section I., Book III.); but $FM = \frac{1}{2}(AB - AC)$. Hence $FM^2 = FR \cdot FI$ (Prop. 8, *Cor.* 5, Section I., Book IV.) $= FK \cdot FN$; ∴ square of tangent from $N = FN \cdot NK$. Again, let GT be the tangent from G to O; then GT^2 = square of tangent from $N + GN^2 = FN \cdot NK + GN^2 = GF^2$. Hence the ⊙ whose centre is G and radius GF will cut the circle O orthogonally; and ∴ that ⊙ will invert the circle O into itself, and the same ⊙ will invert the line BC into Σ; and since BC touches O, their inverses will touch (Prop. 2). Hence Σ touches O, and it is evident that S is the point of contact.

In like manner, if M′ be the point of contact of O′ with BC, and if we join GM′, and let fall the ⊥ HS′ on GM′, S′ will be the point of contact of Σ with O′.

Cor.—The circle on FR as diameter cuts the circles O, O′ orthogonally.

Prop. 11.—Dr. Hart's Extension of Feuerbach's Theorem:—*If the three sides of a plane triangle be replaced by three circles, then the circles touching these, which correspond to the inscribed and escribed circles of a plane triangle, are all touched by another circle.*

Dem.—Let the direct common tangents be denoted, as in Prop. 11, by $\overline{12}$, &c., and the transverse by $\overline{12'}$, &c., and supposing the signs to correspond to a △ whose sides are in order of magnitude a, b, c; then we have, because the side a is touched by the ⊙ 1 on one side, and by the ⊙s 2, 3, 4 on the other side,

$$\overline{12'} \cdot \overline{34} + \overline{14'} \cdot \overline{23} = \overline{13'} \cdot \overline{24};$$

$$\overline{12'} \cdot \overline{34} + \overline{24'} \cdot \overline{13} = \overline{23'} \cdot \overline{14};$$

$$\overline{13'} \cdot \overline{24} + \overline{34'} \cdot \overline{12} = \overline{23'} \cdot \overline{14}.$$

Hence $$\overline{14'} \cdot \overline{23} + \overline{34'} \cdot \overline{12} = \overline{24'} \cdot \overline{13};$$

showing that the four circles are all touched by a circle having the circle 4 on one side, and the other three circles on the other. This proof of Dr. Hart's extension of Feuerbach's theorem was published by me in the *Proceedings of the Royal Irish Academy* in the year 1866.

Prop. 12.—*If two circles* X, Y *be so related that a triangle may be inscribed in* X *and described about* Y, *the inverse of* X *with respect to* Y *is the "Nine-points Circle" of the triangle formed by joining the points of contact on* Y.

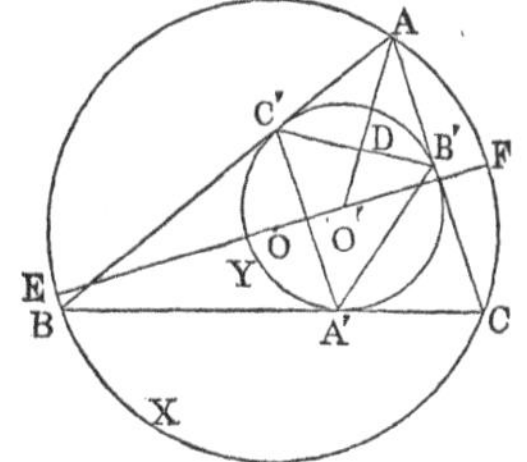

Dem.—Let ABC be the △ inscribed in X and described about Y; and A′B′C′ the △ formed by joining the points of contact on Y.

Let O, O′ be the centres of X and Y. Join O′A, intersecting B′C′ in D; then, evidently, D is the inverse of the point A with respect to Y, and D is the middle point of B′C′. In like manner, the inverses of the points B and C are the middle points C′A′ and A′B′; ∴ the inverse of the ⊙ X, which passes through the points A, B, C with respect to Y, is the ⊙ which passes through the middle points of B′C′, C′A′, A′B′, that is the "Nine-points Circle" of the triangle A′B′C′.

Cor. 1.—If two ⊙s X, Y be so related that a △ inscribed in X may be described about Y, the ⊙ inscribed in the △, formed by joining the points on Y, touches a fixed circle, namely, the inverse of X with respect to Y.

Cor. 2.—In the same case, if tangents be drawn to X at the points A, B, C, forming a new △ A″B″C″, the ⊙ described about A″B″C″ touches a fixed circle.

Cor. 3.—Join OO′, and produce to meet the ⊙ X in the points E and F, and let it meet the inverse of X with respect to Y in the points P and Q; then PQ is the diameter of the "Nine-points Circle" of the △

A′B′C′, and is ∴ = to the radius of Y. Now, let the radii of X and Y be R, r, and let the distance OO′ between their centres be denoted by δ; then we have, because P is the inverse of E, and Q of F,

$$O'P = \frac{r^2}{R + \delta}, \quad O'Q = \frac{r^2}{R - \delta};$$

but $$O'P + O'Q = PQ = r;$$

therefore $$\frac{r^2}{R + \delta} + \frac{r^2}{R - \delta} = r.$$

Hence $$\frac{1}{R + \delta} + \frac{1}{R - \delta} = \frac{1}{r};$$

a result already proved by a different method (see Prop. 11, Section I.).

Prop. 13.—*If a variable chord of a circle subtend a right angle at a fixed point, the locus of its pole is a circle.*

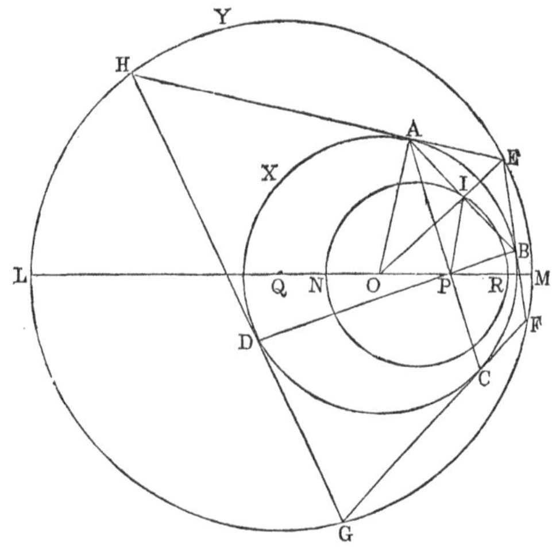

Dem.—Let X be the given circle, AB the variable

chord which subtends a right ∠ at a fixed point P; AE, BE tangents at A and B, then E is the pole of AB: it is required to find the locus of E. Let O be the centre of X. Join OE, intersecting AB in I; then, denoting the radius of X by r, we have $OI^2 + AI^2 = r^2$; but AI = IP, since the ∠ APB is right; ∴ $OI^2 + IP^2 = r^2$; ∴ in the △ OIP there are given the base OP in magnitude and position, and the sum of the squares of OI, IP in magnitude. Hence the locus of the point I is a ⊙ (Prop. 2, *Cor.*, Book II.). Let this be the ⊙ INR. Again, since the ∠ OAE is right, and AI is ⊥ to OE, we have $OI \cdot OE = OA^2 = r^2$. Hence the point E is the inverse of the point I with respect to the ⊙ X; and since the locus of I is a ⊙, the locus of E will be a circle (see Prop. 1).

Prop. 14.—*If two circles, whose radii are* R, r, *and distance between their centres* δ, *be such that a quadrilateral inscribed in one is circumscribed about the other; then*

$$\frac{1}{(R+\delta)^2} + \frac{1}{(R-\delta)^2} = \frac{1}{r^2}.$$

Dem.—Produce AP, BP (see last fig.) to meet the ⊙ X again in the points C and D; then, since the chords AD, DC, CB subtend right ∠s at P, the poles of these chords, viz., the points H, G, F, will be points on the locus of E; then, denoting that locus by Y, we see that the quadrilateral EFGH is inscribed in Y and circumscribed about X. Let Q be the centre of Y; then radius of Y = R, and OQ = δ. Now, since N is a point on the locus of I (see Dem. of last Prop.), $ON^2 + PN^2 = r^2$; but PN = OR; ∴ $ON^2 + OR^2 = r^2$. Again, let OQ produced meet Y in the points L and M; then L and M are the inverses of the points N and R with respect to X. Hence

$$ON \cdot OL = r^2; \text{ that is } ON \cdot (R + \delta) = r^2;$$

therefore $$ON = \frac{r^2}{R+\delta}.$$

In like manner, $OR = \dfrac{r^2}{R - \delta}$;

but we have proved $ON^2 + OR^2 = r^2$;

therefore $$\frac{r^4}{(R + \delta)^2} + \frac{r^4}{(R - \delta)^2} = r^2;$$

or $$\frac{1}{(R + \delta)^2} + \frac{1}{(R - \delta)^2} = \frac{1}{r^2}.$$

This Proposition is an important one in the Theory of Elliptic Functions (see Durége, *Theorie der Elliptischen Functionen*, p. 185). Our proof is as simple and elementary as could be desired. For another proof, by R. F. Davis, M.A., see *Educational Times* (reprint), vol. xxxii.

Prop. 15.—*If* ABC *be a plane triangle,* AD, BE, CF *its perpendiculars,* O *their point of intersection, then the four circles whose centres are* A, B, C, O, *and the squares of whose radii are respectively equal to the rectangles* AO . AD, BO . BE, CO . CF, OA . OD, *are mutually orthogonal.*

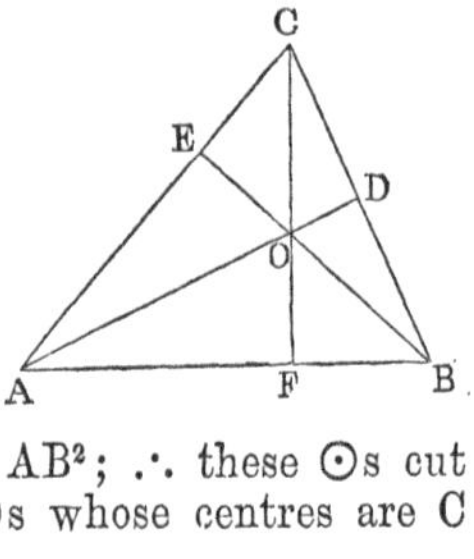

Dem.—AO . AD + BO . BE = AF . AB + BF . BA = AB^2. Hence the sum of the squares of the radii of the ⊙s whose centres are the points A, B = AB^2; ∴ these ⊙s cut orthogonally. Similarly the ⊙s whose centres are C and A cut orthogonally.

Again, let us consider the fourth ⊙, whose centre is the point O, and the square of whose radius is = to the rectangle OA . OD. Now, since OA and OD are measured in opposite directions, they have contrary signs; ∴ the rectangle OA . OD is negative, and the ⊙ has a radius whose square is negative; hence it is imaginary; but, notwithstanding this, it fulfils the condition of intersecting the other ⊙s orthogonally. For AO . AD + OA . OD = AO . AD − AO . OD = AO^2; that is, the sum of the squares of the radii of the circles whose centres are at the points A, O = AO^2. Hence these circles cut orthogonally.

Observation.—In this Demonstration we have made the △ acute-angled, and the imaginary ⊙ is the one whose centre is at the intersection of the ⊥s, and the three others are real; but if the △ had an obtuse angle, the imaginary ⊙ would be the one whose centre is at the obtuse angle.

Prop. 16.—*If four circles be mutually orthogonal, and if any figure be inverted with respect to each of the four circles in succession, the fourth inversion will coincide with the original figure.*

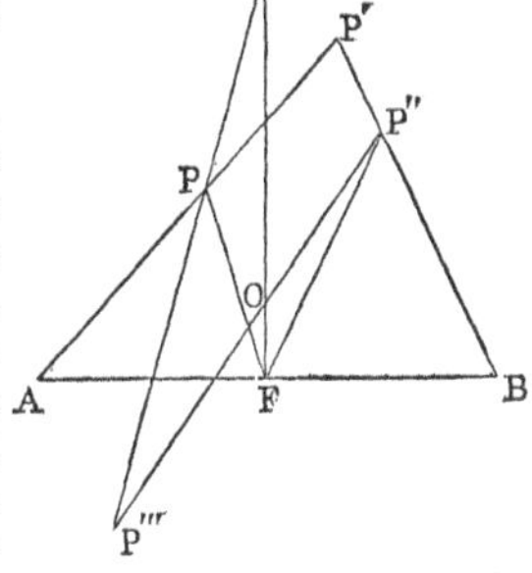

Dem.—It will plainly be sufficient to prove this Proposition for a single point, for the general Proposition will then follow. Let the centres of the four ⊙s be the angular points A, B, C of a △, and O the intersection of its ⊥s: the squares of the radii will be AB . AF, BA . BF, – CO . OF, CF . CO. Now let P be the point we operate on, and let P′ be its inverse with respect to the ⊙ A, and P″ the inverse of P′ with respect to the ⊙ B. Join P″O and CP meeting in P‴. Now, since P′ is the inverse of P with respect to the ⊙ A, the square of whose radius is AB . AF, we have AB . AF = AP . AP′; ∴ the △ AFP is equiangular to the △ AP′B; ∴ ∠ AFP = AP′B: in like manner the ∠ BFP″ = AP′B, ∴ the △s AFP, BP″F are equiangular, ∴ rectangle AF . FB = PF . FP″. Again, because O is the intersection of the ⊥s of the △ ABC, AF . FB = CF . OF. Hence CF . OF = PF . FP″, and the ∠s CFP and OFP″ are equal, since the ∠s AFP and BFP″ are equal; ∴ the △s P″FO and CFP are equiangular, and the ∠s OP″F and PCF are equal; hence the four points C, P″, F, P‴ are concyclic;

∴ rectangle $OP'' . OP''' =$ rectangle $OC . OF$; the point P''' is the inverse of P'' with respect to the ⊙ whose centre is O, and the square of whose radius is the negative quantity $OC . OF$. Again, the ∠ $OFP = P''FO = OP'''P$, ∴ the four points O, F, P''', P are concyclic; ∴ $CP . CP''' = CO . CF$, and the point P is the inverse of P''' with respect to the ⊙ whose centre is C, and the square of whose radius is the rectangle $CF . CO$. Hence the Proposition is proved.

The foregoing theorem is important in the Theory of Elliptic Functions, as on it depends the reduction of the rectification of Bicircular and Sphero-Quartics to Elliptic Integrals (see *Phil. Trans.*, vol. 167, Part ii., "On a New Form of Tangential Equation").

The following elegant proof, which has been communicated to the author by W. S. M'Cay, F.T.C.D., depends on the principle (Miscellaneous Exercises, No. 60), that a circle and two inverse points invert into a circle and two inverse points.

Invert the four orthogonal circles from an intersection of two of them and we get a circle (radius R), two rectangular diameters, and an imaginary concentric circle (radius $R\sqrt{-1}$). Successive inversions with respect to these two circles turn P into Q ($OP = - OQ$); and successive reflexions in the two diameters bring Q back to P.

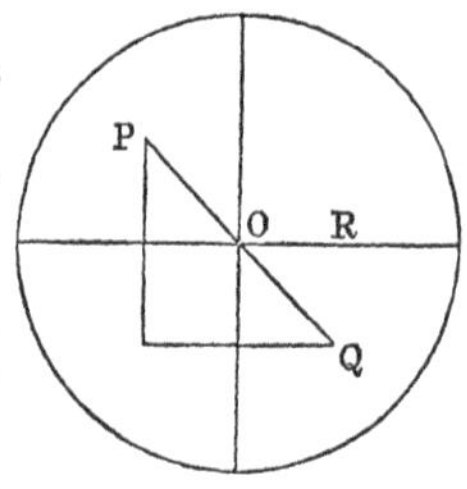

This theorem can be extended to surfaces, thus: "If five spheres be mutually orthogonal, and if any surface be inverted with respect to each of the five spheres in succession, the fifth inversion will coincide with the original surface."

SECTION V.

Coaxal Circles.

In Book III., Section I., Prop. 24, we have proved the following theorem:—

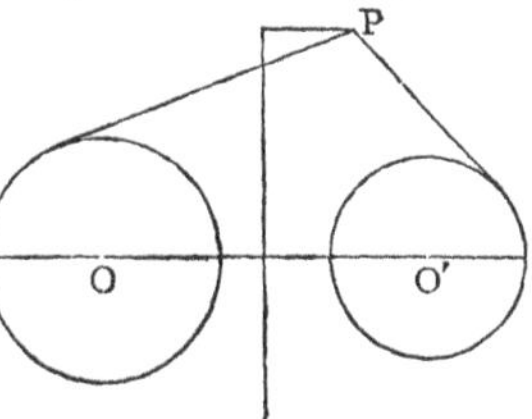

"*If from any point* P *tangents be drawn to two circles, the difference of their squares is equal twice the rectangle contained by the perpendicular let fall from* P *on the radical axis and the distance between their centres.*"

The following special cases of this theorem are deserving of notice:—

(1). Let P be on the circumference of one of the circles, and we have—*If from any point* P *in the circumference of one circle a tangent be drawn to another circle, the square of the tangent is equal twice the rectangle contained by the distance between their centres and the perpendicular from* P *on the radical axis.*

(2). Let the circle to which the tangent is drawn be one of the limiting points, *then the square of the line drawn from one of the limiting points to any point of a circle of a coaxal system varies as the perpendicular from that point on the radical axis.*

(3). *If* X, Y, Z *be three coaxal circles, the tangents drawn from any point of* Z *to* X *and* Y *are in a given ratio.*

(4). *If tangents drawn from a variable point* P *to two given circles* X *and* Y *have a given ratio, the locus of* P *is a circle coaxal with* X *and* Y.

(5). *The circle of similitude of two given circles is coaxal with the two circles.*

(6). *If* A *and* B *be the points of contact, upon two circles* X *and* Y, *of tangents drawn from any point of their circle of similitude, then the tangent from* A *to* Y *is equal to the tangent from* B *to* X.

Prop. 2.—*Two circles being given, it is required to describe a system of circles coaxal with them.*

Con.—If the circles have real points of intersection, the problem is solved by describing circles through these points and any third point taken arbitrarily.

If the given circles have not real points of intersection, we proceed as follows:—

Let X and Y be the given ⊙s, P and Q their centres: draw AB, the radical axis of X and Y, intersecting PQ in O: from O draw two tangents OC, OD

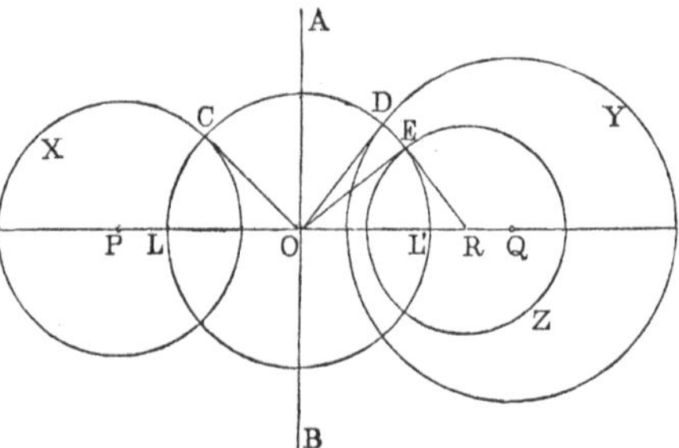

to X and Y; then OC = OD, and the ⊙ described with O as centre and OD as radius will cut the two ⊙s X and Y orthogonally. Now take any point E in this orthogonal ⊙, and draw the tangent ER meeting the line PQ in R: from R as centre, and RE as radius, describe a ⊙ Z; then Z will be coaxal with X and Y. For the line ER being a tangent to the ⊙ CDE, the ∠ OER is right, ∴ OE is a tangent to Z; and since OD = OE, the tangents from O to the ⊙s Y and Z are equal: hence OA is the radical axis of Y and Z; ∴ the three ⊙s X, Y, Z are coaxal. In like manner, we can get another circle coaxal with X and Y by taking any other point in the ⊙ CDE, and drawing a tangent, and repeating the same construction as with the ⊙ Z. In this way we evidently get two infinite systems of circles coaxal with X and Y, namely, one system at each side of the radical axis. The smallest circle of each system is a point, namely, the point at each side of the radical axis in which the line joining the centres of X and Y cuts the ⊙ CDE. These are the limiting points, and in this point of view we see that each limiting point is to be regarded as an infinitely small circle. The two infinite systems of circles are to be regarded as one coaxal system, the circles of which range from infinitely large to infinitely small—the radical axis being the infinitely large circle, and the limiting points the infinitely small.

Cor. 1.—No circle of a system with real limiting points can have its centre between the limiting points.

Cor. 2.—The centres of the circles of a coaxal system are collinear.

Cor. 3.—The circle described on the distance between the limiting points as diameter cuts all the circles of the system orthogonally.

Cor. 4.—Every circle passing through the limiting points cuts all the circles of the system orthogonally.

Cor. 5.—The limiting points are inverse points with respect to each circle of the system.

Cor. 6.—The polar of either limiting point, with respect to every circle of the system, passes through the other, and is perpendicular to the line of collinearity of their centres.

Prop. 8.—*If two circles* X *and* Y *cut orthogonally, the polar with respect to* X *of any point* A *in* Y *passes through* B, *the point diametrically opposite to* A.

This is Prop. 26, Book III., Section I. The following are important deductions:—

Cor. 1.—The circle described on the line joining a point A to any point B in its polar, with respect to a given circle, cuts that circle orthogonally.

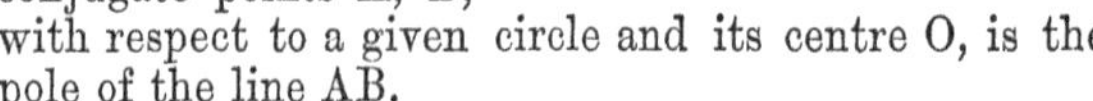

Cor. 2.—The intersection of the ⊥s of the △ formed by a pair of conjugate points A, B, with respect to a given circle and its centre O, is the pole of the line AB.

Cor. 3.—The polars of any point A with respect to a coaxal system are concurrent. For, through A and through the limiting points describe a ⊙ : this (*Cor.* 4, Prop. 2) will cut all the ⊙s orthogonally, and the polars of A with respect to all the ⊙s of the system will pass through the point diametrically opposite to A on this orthogonal ⊙ ; hence they are concurrent.

Cor. 4.—If the polars of a variable point with respect to three given ⊙s be concurrent, the locus of the point is the ⊙ which cuts the three given ⊙s orthogonally.

Prop. 4.—*If* X_1, X_2, X_3, &c., *be a system of coaxal circles, and if* Y *be any other circle, then the radical axes of the pairs of circles* X_1, Y ; X_2, Y ; X_3, Y, &c., *are concurrent.*

Dem.—The two first meet on the radical axis of X_1, X_2 ; the second and third on the radical axis of X_2, X_3 ; but this, by hypothesis, is the radical axis of X_1, X_2 ; hence the Proposition is evident.

Prop. 5.—*If two circles cut two other circles orthogonally, the radical axis of either pair is the line joining the centres of the other pair.*

Dem.—Let X, Y be one pair cutting W, V, the other pair, orthogonally; then, since X cuts W and V orthogonally, the tangents drawn from the centre of X to W and V are equal; hence the radical axis of W and V passes through the centre of X. In like manner the radical axis of W and V passes through the centre of Y; ∴ the line joining the centres of the ⊙s X and Y is the radical axis of the ⊙s W and V. In the same way it can be shown that the line joining the centres of W and V is the radical axis of X and Y.

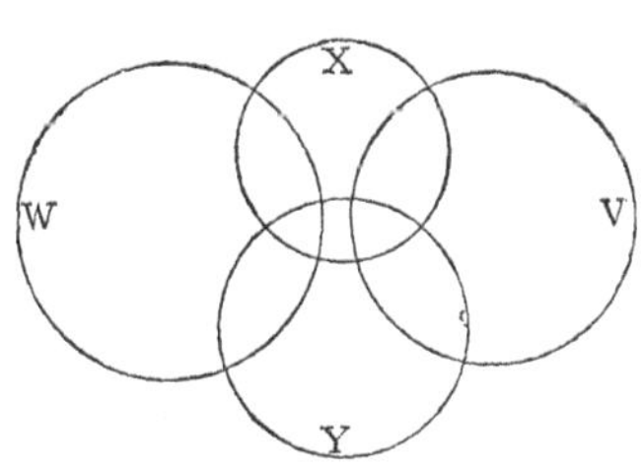

Cor. 1.—If one pair of the ⊙s, such as W and V, do not intersect, the other pair, X, Y, will intersect, because they must pass through the limiting points of W and V.

Cor. 2.—Coaxal ⊙s may be divided into two classes—one system not intersecting each other in real points, but having real limiting points; the other system intersecting in real points, and having imaginary limiting points.

Cor. 3.—If a system of circles be cut orthogonally by two circles they are coaxal.

Cor. 4.—If four circles be mutually orthogonal, the six lines joining their centres, two by two, are also their radical axes, taken two by two.

Prop. 6.—*If a system of concentric circles be inverted from any arbitrary point, the inverse circles will form a coaxal system.*

Dem.—Let O be the centre of inversion, and P the common centre of the concentric system. Through P draw any two lines: these lines will cut the concentric system orthogonally, and therefore their inverses, which will be two circles passing through the point O and through the inverse of P, will cut the inverse of the concentric system orthogonally; hence the inverse of the concentric system will be a coaxal system (Prop. 5, *Cor.* 3).

Cor. 1.—The limiting points will be the centre of

inversion, and the inverse of the common centre of the original system.

Cor. 2.—If a variable circle touch two concentric circles, it will cut any other circle concentric with them at a constant angle. Hence, by inversion, if a variable circle touch two circles of a coaxal system, it will cut any other circle of the system at a constant angle.

Cor. 3.—If a variable circle touch two fixed circles, its radius has a constant ratio to the perpendicular from its centre on the radical axis of the two circles, for it cuts the radical axis at a constant angle.

Cor. 4.—The inverse of a system of concurrent lines is a system of coaxal ⊙s intersecting in two real points.

Cor. 5.—If a system of coaxal circles having real limiting points be inverted from either limiting point, they will invert into a concentric system of circles.

Cor. 6.—If a coaxal system of either species be inverted from any arbitrary point, it inverts into another system of the same species.

Prop. 7.—*If a variable circle touch two fixed circles, its radius has a constant ratio to the perpendicular from its centre on the radical axis.*

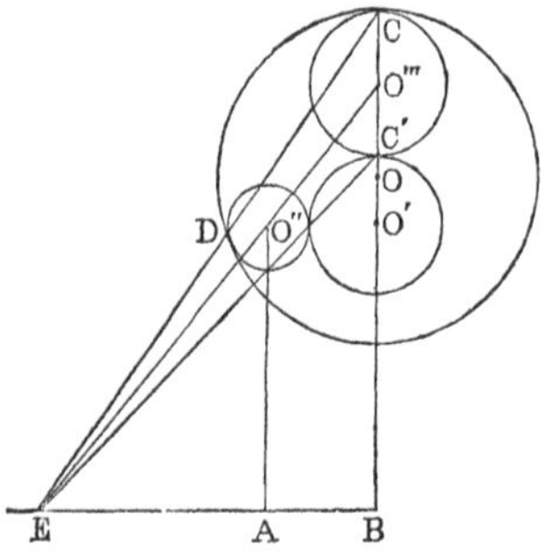

Dem.—This is *Cor.* 3 of the last Proposition; but it is true universally, and not only as proved there for the case where the ⊙ cuts the radical axis. On account of its importance we give an independent proof here. Let the centres of the fixed ⊙s be O, O′, and that of the variable ⊙ O″. Join OO′, and produce it to meet the fixed ⊙s in the points C, C′: upon CC′ describe a ⊙: let O‴ be its centre: let fall the ⊥s O″A, O‴B on the radical axis: let D be the point of contact of O″ with O; then the lines CD and O‴O″ will meet in the centre of similitude of the ⊙s O'', O'''; but this centre is a point on the radical axis of the circles O, O′ (see Prop. 4, Section II.). Hence the point E is on the radical axis, and, by similar triangles,

$$O''A : O'''B :: O''E : O'''E :: \text{radius of } O'' : \text{radius of } O''',$$

$$\therefore \text{radius of } O'' : O''A :: \text{radius of } O''' : O'''B;$$

but the two last terms of this proportion are constant,

$$\therefore \text{radius of } O'' : O''A \text{ in a constant ratio.}$$

Prop. 8.—*If a chord of one circle be a tangent to another, the angle which the chord subtends at either limiting point is bisected by the line drawn from that limiting point to the point of contact.*

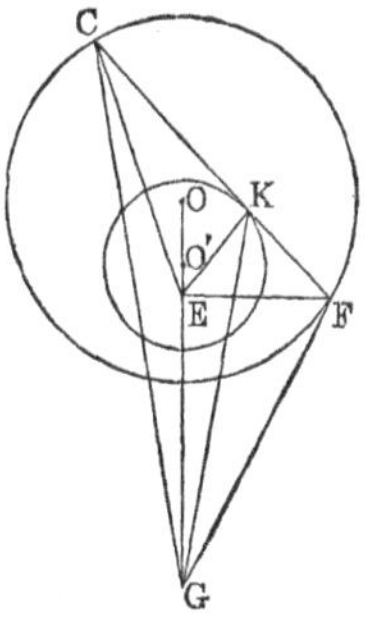

Let CF be the chord, K the point of contact, E one of the limiting points: the angle CEF is bisected by EK. For since the limiting point E is coaxal with the circles O, O′ we have, by Prop. I. (3),

$$CE : CK :: FE : FK;$$

$$\therefore EC : EF :: KC : KF.$$

Hence the angle CEF is bisected (VI. iii).

In like manner, if G be the other limiting point, the angle CGF is bisected by GK.

Cor. 1.—If the circles were external to each other, and the figure constructed, it would be found that the angles bisected would be the supplements of the angles CEF, CGF.

Cor. 2.—If a common tangent be drawn to two circles, lines drawn from the points of contact to either limiting point are perpendicular to each other; for they are the internal and external bisectors of an angle.

Cor. 3.—If three circles be coaxal, a common tangent to two of them will intersect the third in points which are harmonic conjugates to the points of contact; for the pencil from either limiting point will be a harmonic pencil.

Cor. 4.—If a circle be described about the triangle CEF, its envelope will be a circle concentric with the circle whose centre is O; that is, with the circle whose chord is CF.

(*When a line or circle moves according to any given law, the curve which it touches in all its positions is called its envelope.*)

Produce EK till it meets the circumference in D; then because the ∠ CEF is bisected by ED, the arc CDF is bisected in D; hence the line OG, which joins the centres of the circles, passes through D and is ⊥ to CF; ∴ O′K is ∥ to OD; ∴ O′K : OD : : EO′ : EO; hence the ratio of O′K : OD is given; but O′K is given; therefore OD is given, and the ⊙ whose centre is O and radius OD is given in position, and the ⊙ CEF touches it in D; hence the Proposition is proved.

Prop. 9.—*If a system of coaxal circles have two real points of intersection, all lines drawn through either point are divided proportionally by the circles.*

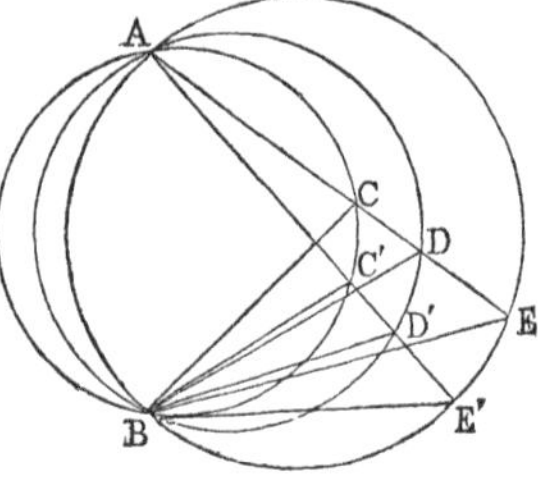

Let A, B be the points of intersection of the coaxal system: through A draw two lines intersecting the circles again in the two systems of points C, D, E; C′, D′, E′; then

CD : DE : : C′D′ : D′E′.

Dem.—Join the points C, D, E, C′, D′, E′ to B; then the △s BCD, BC′D′ are evidently equiangular, as are also the triangles BDE, BD′E′; hence

CD : DB : : C′D′ : D′B;

DB : DE : : D′B : D′E′;

therefore, *ex aequali*,

CD : DE : : C′D′ : D′E′.

Cor. 1.—If two lines be divided proportionally, the circles passing through their point of intersection and through pairs of homologous points are coaxal.

Cor. 2.—If from the point B perpendiculars be drawn to the lines joining homologous points, the feet of these perpendiculars are collinear. For each lies on the line joining the feet of the perpendiculars from B on the lines AC, AC′.

Cor. 3.—The circles described about the triangles formed by the lines joining any three pairs of homologous points all pass through B.

Cor. 4.—The intersection of the perpendiculars of all the triangles formed by the lines joining homologous points are collinear.

Cor. 5.—Any two lines joining homologous points are divided proportionally by the remaining lines of the system.

Prop. 10.—*To describe a circle touching three given circles.*

Analysis.—Let X, Y, Z be the three given ⊙s, ABC, A′B′C′ two ⊙s which it is required to describe touching the three given ⊙s; then, by *Cor.* 2, Prop. 4, Section IV., the ⊙ DEF, which cuts X, Y, Z orthogonally, will be the ⊙ of inversion of ABC, A′B′C′, and the three ⊙s ABC, DEF, A′B′C′ will be coaxal (*Cor.* 2, Prop. 1, Section IV.).

Now, consider the ⊙ X, and the three ⊙s ABC, DEF, A′B′C′; the radical axes of X and these ⊙s are concurrent (Prop. 4); but two of the radical axes are tangents at A, A′, and the third is the common chord of X and the orthogonal ⊙ DEF; let P be their point of concurrence. Again, from Prop. 4, Section II., it follows that the axis of similitude of X, Y, Z is the radical axis of the ⊙s ABC, A′B′C′; but since PA = PA′, being tangents to X, the point P is on this radical axis. Hence P is the point of intersection of two given lines, namely, the axis of similitude of X, Y, Z, and the chord common to X and the orthogonal ⊙ DEF; ∴ P is a given point; hence A, A′, the points of contact of the tangents from P to X, are given. Similarly, the points

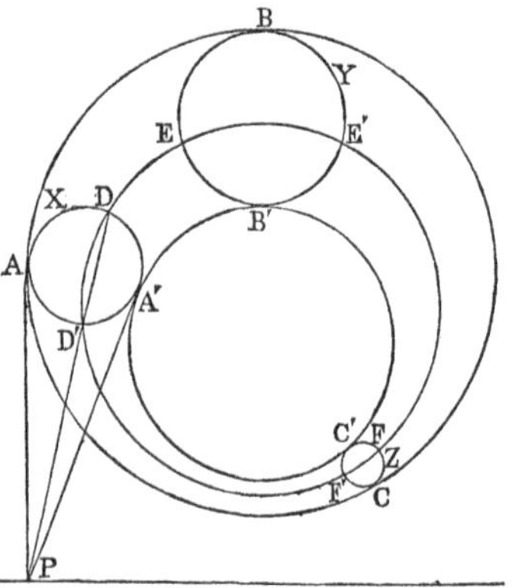

B, B′; C, C′ are given points. And we have the following construction, viz.: *Describe the orthogonal circle of* X, Y, Z, *and draw the three chords of intersection of this circle with* X, Y, Z *respectively; and from the points where these chords meet the axis of similitude of* X, Y, Z *draw pairs of tangents to* X, Y, Z; *then the two circles described through these six points of contact will be tangential to* X, Y, Z.

Cor. 1.—Since there are four axes of similitude of X, Y, Z, we shall have eight circles tangential to X, Y, Z.

Cor. 2.—If we suppose one of the circles to reduce to a point, we have the problem: "*To describe a circle touching two given circles, and passing through a given point.*" And if two of the circles reduce to points, we have the problem: "*To describe a circle touching a given circle, and passing through two given points.*"

The foregoing construction holds for each case, the first of which admits of four solutions, and the second of two.

Cor. 3.—Similarly, we may suppose one of the circles to open out into a line, and we have the problem: "*To describe a circle touching a line and two given circles*"; and if two circles open out into lines, the problem: "*To describe a circle touching two given lines and a circle.*" The foregoing construction extends to these cases also, and like observations apply to the remaining cases, namely, when one of the circles reduces to a point, and one opens out into a line, &c. Since our construction embraces all cases, except where the three circles become three points or open out into three lines, it would appear to be the most general construction yet given for the solution of this celebrated problem.

Another Method—**Analysis.**—Let O, O′, O″ be the centres of the ⊙s X, Y, Z, and let AR, BR be the radical axis of the pairs of ⊙s XY, YZ, respectively, and let O‴ be the centre of the required ⊙ W: from O‴ let fall the ⊥s O‴A, O‴B; join R to C, the point of contact of W with Z, and produce it to meet O″D drawn ∥ to O‴R. Now, because W touches the ⊙s X, Y, its radius O‴C has a given ratio to O‴A (Prop. 7). Similarly, O‴C has a given ratio O‴B; ∴ O‴A has a given ratio to O‴B; hence the line O‴R is given in position, and the ratio of O‴R : O‴B is given; ∴ the ratio of O‴R : O‴C is given; hence the ratio of O″D : O″C is given; ∴ D is a given point and R is a given point; ∴ the line RD is given in position; hence C is a given point. Similarly, the other points of contact are given.

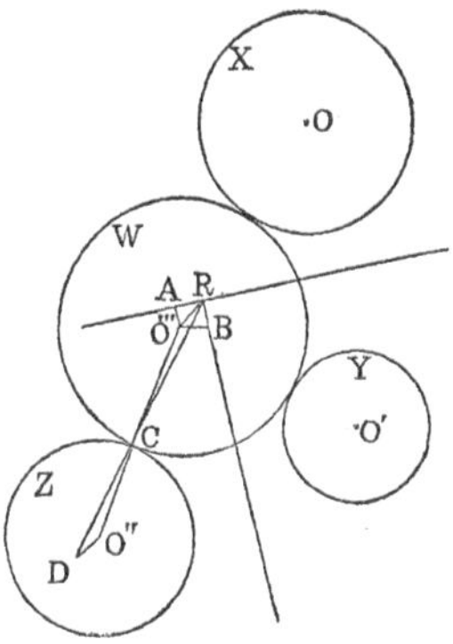

Observation.—This method, though arrived at by the theory of coaxal circles, is virtually the same as Newton's 16th Lemma. It is, however, somewhat simpler, as it does not employ conic sections, as is done in the *Principia*. When I discovered it several years ago, I was not aware to what an extent I had been anticipated.

Prop. 11.—*If* X, Y *be two circles*, AB, A′B′ *two chords of* X *which are tangents to* Y; *then if the perpendiculars from* A, A′ *on the radical axis be denoted by* p, π, *and the perpendiculars from* B, B′ *by* p', π',

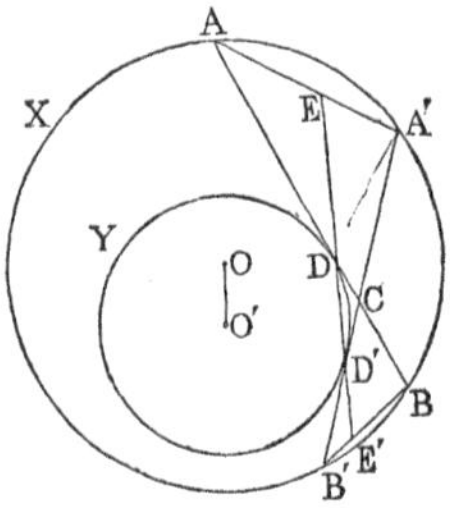

$$AA' : BB' :: \sqrt{p} + \sqrt{\pi} : \sqrt{p'} + \sqrt{\pi'}.$$

Dem.—Let O, O′ be the centres of the circles; then, by (1), Prop. 1,

$$AD = \sqrt{2 \cdot OO' \cdot p}, \qquad A'D' = \sqrt{2 \cdot OO' \cdot \pi};$$

$$\therefore \; AD + A'D' = \sqrt{2 \cdot OO'}\,\{\sqrt{p} + \sqrt{\pi}\}.$$

But AD + A′D′ is easily seen to be = AC + A′C;

$$\therefore \; AC + A'C = \sqrt{2 \cdot OO'}\,\{\sqrt{p} + \sqrt{\pi}\}.$$

In like manner,

$$BC + B'C = \sqrt{2 \cdot OO'}\,\{\sqrt{p'} + \sqrt{\pi'}\}.$$

Hence,

$$AC + A'C : BC + B'C :: \sqrt{p} + \sqrt{\pi} : \sqrt{p'} + \sqrt{\pi'}.$$

Now, since the triangles AA′C, BB′C are equiangular, we have

$$AC + A'C : BC + B'C :: AA' : BB';$$

$$\therefore \; AA' : BB' :: \sqrt{p} + \sqrt{\pi} : \sqrt{p'} + \sqrt{\pi'}.$$

This theorem is very important, besides leading to an immediate proof of *Poncelet's Theorem*. If we suppose the chords AB, A'B' to be indefinitely near, we can infer from it a remarkable property of the motion of a particle in a vertical circle, and also a method of representing the amplitude of Elliptic Integrals of the First kind by coaxal circles.*

Prop. 12.—PONCELET'S THEOREM.—*If a variable polygon of any number of sides be inscribed in a circle of a coaxal system, and if all the sides but one in every position touch fixed circles of the system, that one also in every position touches another fixed circle of the system.*

It will be sufficient to prove this Theorem for the case of a triangle, because from this simple case it is easy to see that the Theorem for a polygon of any number of sides is an immediate consequence.

Let ABC be a △ inscribed in a ⊙ of the system, A'B'C' another position of the △, and let the sides AB, A'B' be tangents to one ⊙ of the system, BC, B'C' tangents to another ⊙; then it is required to prove that CA, C'A' will be tangents to a third ⊙ of the system.

Dem.—Let the perpendiculars from A, B, C on the radical axis be denoted by p, p', p'', and the perpendiculars from A', B', C' by π, π', π''; then, by Prop. 11, we have

$$AA' : BB' :: \sqrt{p} + \sqrt{\pi} : \sqrt{p'} + \sqrt{\pi'},$$

and $$BB' : CC' :: \sqrt{p'} + \sqrt{\pi'} : \sqrt{p''} + \sqrt{\pi''};$$

$$\therefore AA' : CC' :: \sqrt{p} + \sqrt{\pi} : \sqrt{p''} + \sqrt{\pi''}.$$

Hence AC, A'C are tangents to another circle of the system.

The foregoing proof of this celebrated theorem was given by me in 1858 in a letter to the Rev. R. Townsend, F.T.C.D. It is virtually the same as Dr. Hart's proof, published in 1857 in the *Quarterly Journal of Mathematics*, of which I was not aware at the time.

* The method of representing the amplitude of Elliptic Integrals by coaxal circles was first given by Jacobi, Crelle's *Journal*, Band. III. Theorem 11 affords a very simple proof of this application. See *Educational Times*, Vol. III., Reprint, page 42.

Dr. Hart's Proof.—This proof depends on the following Lemma (see fig., Prop. 11):—If a quadrilateral AA′BB′ be inscribed in a circle X, and if the diagonals AB, A′B′ touch a circle Y of a system coaxal with X, then the sides A, A′ touch another circle of the same system, and the four points of contact D, D′, E, E′ are collinear.

This proposition is evident from the similar triangles AED, B′E′D′, and the similar triangles EA′D′, E′BD; and the equality of the ratios AE : AD, B′E′ : B′D′, A′E : A′D, BE : BD.

The first part of this theorem also follows at once from Prop. 11.

Now, to prove Poncelet's theorem:—Let ABC, A′B′C′ be two positions of the variable △, and let, as before, AB, A′B′ be tangents to one ⊙ of the system, BC, B′C′ tangents to another ⊙; then CA, C′A′ shall be tangents to a third ⊙ of the system. For, join AA′, BB′, CC′. Then, since AB, A′B′ are tangents to a ⊙ of the system, AA′, BB′ are, by the lemma, tangents to another ⊙ of the system; and since BC, B′C′ are tangents to a ⊙ of the system, BB′, CC′ are tangents to a ⊙ of the system; ∴ AA′, BB′, CC′ are tangents to a ⊙ of the system; and since AA′, CC′ touch a ⊙ of the system, by the lemma, AC, A′C′ touch a ⊙ of the system; hence the Proposition is proved, and we see that the two proofs are substantially identical.

SECTION VI.

Theory of Anharmonic Section.

Def.—*A system of four collinear points* A, B, C, D *make, as is known, six segments; these may be arranged in three pairs, each containing the four letters—thus,*

AB, CD; BC, AD; CA, BD.

Where the last letter in each couple is D, *and the first segments in the three couples are respectively* AB, BC, CA, *exactly corresponding to the sides of a triangle* ABC, *taken in order. Now, if we take the rectangles formed by these three pairs of segments, the six quotients obtained by dividing each rectangle by the two remaining ones are called the six anharmonic ratios of the four points* A, B, C, D. *Thus these six functions are*

$$\frac{AB \cdot CD}{BC \cdot AD}, \quad \frac{BC \cdot AD}{CA \cdot BD}, \quad \frac{CA \cdot BD}{AB \cdot CD};$$

and their reciprocals

$$\frac{BC \cdot AD}{AB \cdot CD}, \quad \frac{CA \cdot BD}{BC \cdot AD}, \quad \frac{AB \cdot CD}{CA \cdot BD}.$$

It is usual to call any one of these six functions the anharmonic ratio of the four points A, B, C, D.

Prop. 1.—*If* (O . ABCD) *be a pencil of four rays passing through the four points* A, B, C, D; *and if through any of these points* B *we draw a line parallel to a ray passing through any of the other points, and cutting the two remaining rays in the points* M, N, *the six anharmonic ratios of* A, B, C, D *can be expressed in terms of the ratios of the segments* MB, BN, NM.

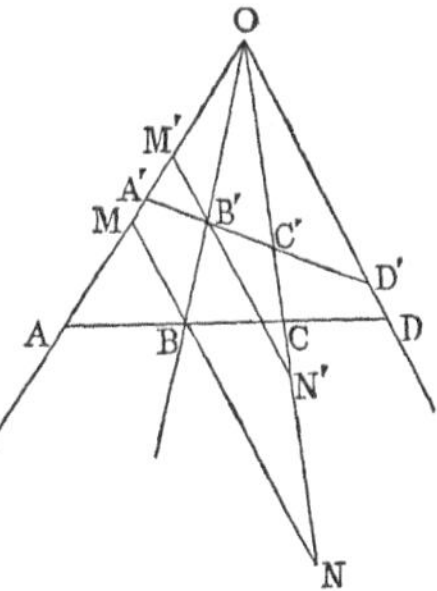

Dem.—From similar triangles,

$$\frac{MB}{OD} = \frac{AB}{AD},$$

and

$$\frac{OD}{BN} = \frac{CD}{BC}.$$

Hence, $$\frac{MB}{BN} = \frac{AB \,.\, CD}{BC \,.\, AD};$$

therefore $$MB : BN :: AB \,.\, CD : BC \,.\, AD.$$

Componendo—

$$MN : BN :: AB \,.\, CD + BC \,.\, AD : BC \,.\, AD;$$

$$\therefore\ MN : BN :: AC \,.\, BD : BC \,.\, AD;$$

$$\therefore\ MB : BN : NM :: AB \,.\, CD : BC \,.\, AD : CA \,.\, BD.$$

Prop. 2.—*If a pencil of four rays be cut by two transversals* ABCD, A′B′C′D′, *then* (see last fig.) *any of the anharmonic ratios of the points* A, B, C, D *is equal to the correspondiug ratio for the points* A′, B′, C′, D′.

Dem.—Through the points B, B′ draw MN, M′N′ parallel to OD; then (Section I., Prop. 3) we have

$$MB : BN :: M'B' : B'N';$$

therefore $$\frac{AB \,.\, CD}{BC \,.\, AD} = \frac{A'B' \,.\, C'D'}{B'C' \,.\, A'D'}.$$

Cor. 1.—We may suppose the rays of the pencil produced through the vertex, and the transversal to cut any of the rays produced without altering the anharmonic ratio.

Def.—*The anharmonic ratio of the four points on any transversal cutting a pencil being constant, it is called the anharmonic ratio of the pencil.*

Cor. 2.—If two pencils have equal anharmonic ratios and a common vertex; and if three rays of one pencil be the production of three rays of the other, then the fourth ray of one is the production of the fourth ray of the other.

Cor. 3.—If two pencils have a common transversal, they are equal; that is, they have equal anharmonic ratios.

Cor. 4.—If A, B, C, D be four points in the circumference of a circle, and E and F any two other points also in the circumference, then the pencil (E . ABCD) = (F . ABCD). This is evident, since the pencils have equal angles.

Cor. 5.—If through the middle point O of any chord AB of a circle two other chords CE and DF be drawn, and if the lines ED and CF joining their extremities intersect AB in G and H, then OG = OH.

Dem.—The pencil (E . ADCB) = (F . ADCB; therefore the anharmonic ratio of the points A, G, O, B = the anharmonic ratio of the points A, O, H, B; and since AO = OB, OG = OH.

Def.—*The anharmonic ratio of the cyclic pencil* (E . ABCD) *is called the anharmonic ratio of the four cyclic points* A, B, C, D.

Prop. 3.—*The anharmonic ratio of four concyclic points can be expressed in terms of the chords joining these four points.*

Dem. (see fig., Prop. 9, Section IV.)—The anharmonic ratio of the pencil (O . ABCD) is AC . BD : AB . CD; and this, by Prop. 9, Section IV. = A′C′ . B′D′ : A′B′ . C′D′; but the pencil (O . ABCD) = the pencil (O . A′B′C′D′) = the anharmonic ratio of the points A′, B′, C′, D′. Hence the Proposition is proved.

Cor. 1.—The six functions formed, as in Def. 1, with the six chords joining the four concyclic points A′, B′, C′, D′, are the six anharmonic ratios of these points.

Cor. 2.—If two triangles CAB, C′A′B′ be inscribed in a circle, any two sides, viz., one from each triangle, are divided equianharmonically by the four remaining sides. For, let the sides be AB, A′B′; then the pencils (C . A′BAB′), (C′ . A′BAB′) are equal (*Cor.* 4, Prop. 2).

Prop. 4.—Pascal's Theorem.—*If a hexagon be inscribed in a circle, the intersections of opposite sides* *viz.*, 1*st and* 4*th*, 2*nd and* 5*th*, 3*rd and* 6*th*, *are collinear.*

Let ABCDEFA be the hexagon. The points L, N, M are collinear.

Dem.—Join EN. Then the pencil (N . FMCE) = the pencil (C . FBDE), because they have a common transversal EF (*Cor.* 3, Prop. 2.) In like manner, the pencil (A . FBDE) = (N . ALDE); but (A.FBDE) = C.FBDE) (Prop. 2, *Cor.* 4). Hence the pencils (N . FMCE), (N . ALDE) are equal; and therefore (*Cor.* 2, Prop. 2) the points L, N, M are collinear.

Cor. 1.—With six points on the circumference of a circle, sixty hexagons can be formed. For, starting with any point, say A, we could go from A to one of the remaining points in five ways. Suppose we select B, then we could go from B to a third point in four different ways, and so on; hence it is evident that we could join A to another point, and that again to another, and so on, and finally return to A in $5 \times 4 \times 3 \times 2 \times 1$ different ways. Hence we shall have that number of hexagons; but each is evidently counted twice, and we shall therefore have half the number, that is, sixty distinct hexagons.

Cor. 2.—Pascal's Theorem holds for each of the sixty hexagons.

Cor. 3.—Pascal's Theorem holds for six points,

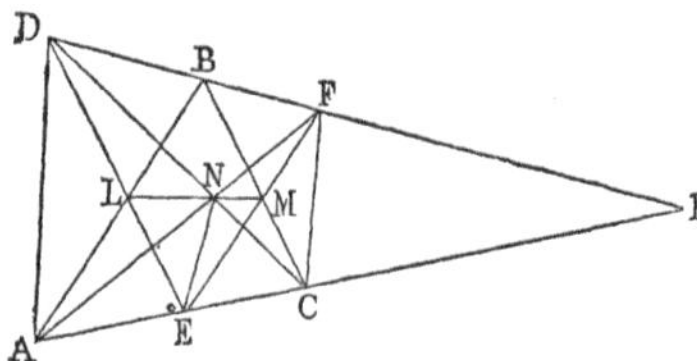

which are, three by three, on two lines. Thus, let the

two triads of points be A, E, C, D, B, F, and the proof of the Proposition can be applied, word for word, except that the pencil (A . FBDE) is equal to the pencil (C . FBDE), for a different reason, viz., they have a common transversal.

Prop. 5.—*If two equal pencils have a common ray, the intersections of the remaining three homologous pairs of rays are collinear.*

Let the pencils be (O . O′ABC), (O′ . OABC), having the common ray OO′; then, if possible, let the line joining the points A and C intersect the rays OB, O′B in different points B′, B″; then, since the pencils are equal, the anharmonic ratio of the points D, A, B′, C equal the anharmonic ratio of the points D, A, B″, C, which is impossible. Hence the points A, B, C must be collinear.

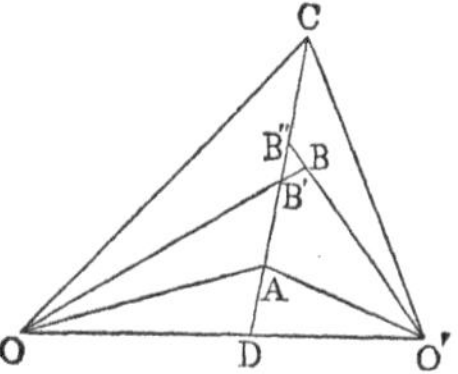

Cor. 1.—If A, B, C; A′, B′, C′ be two triads of points on two lines intersecting in O, and if the anharmonic ratio (OABC) = (OA′B′C′), the three lines AA′, BB′, CC′ are concurrent. For, let AA′, BB′, intersect in D; join CD, intersecting OA′ in E; then the anharmonic ratio (OA′B′E) = (OABC) = (OA′B′C′) by hypothesis; therefore the point E coincides with C′. Hence the Proposition is proved.

Cor. 2.—If two △s ABC, A′B′C′ have lines joining corresponding vertices concurrent, the intersections of corresponding sides must be collinear. For, join P, the point of intersection of the sides BC, B′C′,

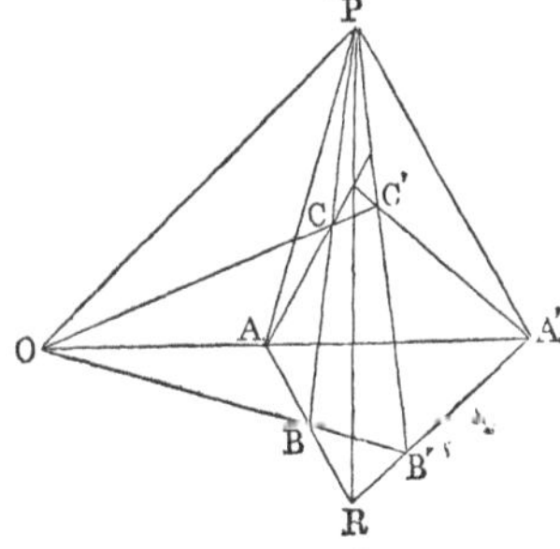

to O, the centre of perspective; then each of the pencils (A . PCA′B), (A′ . PC′AB′) is equal to the pencil (O . PCAB); hence they are equal to one another, and they have the ray AA′ common. Hence the intersections of the three corresponding pairs of rays AC, A′C′, AP, A′P, AB, A′B′, are collinear.

Cor. 3.—If two vertices of a variable △ ABC move on fixed right lines LM, LN, and if the three sides pass through three fixed collinear points O, P, Q, the locus of the third vertex is a right line.

Let the side AB pass through O, BC through P, CA through Q, and let A′B′C′ be another position of the △; then the two △s AA′Q, BB′Q, have the lines joining their corresponding vertices concurrent; hence the intersections of the corresponding sides are collinear. Hence the Proposition is proved.

Prop. 6.—*If on a right line* OX *three pairs of points* A, A′; B, B′; C, C′ *be taken, such that the three rectangles* OA . OA′, OB . OB′, OC . OC′, *are each equal to a constant, say* k^2, *then the anharmonic ratio of any four of the six points is equal to the anharmonic ratio of their four conjugates.*

Dem.—Erect OY at right ∠s to OX, and make OY = k; join AY, A′Y, BY, B′Y, CY, C′Y. Now, by hypothesis, OA . OA′ = OY^2; ∴ the ⊙ described about the △AA′Y touches OY at Y; ∴ the ∠ OYA = OA′Y. In like manner, the ∠ OYB = OB′Y; hence the ∠ AYB

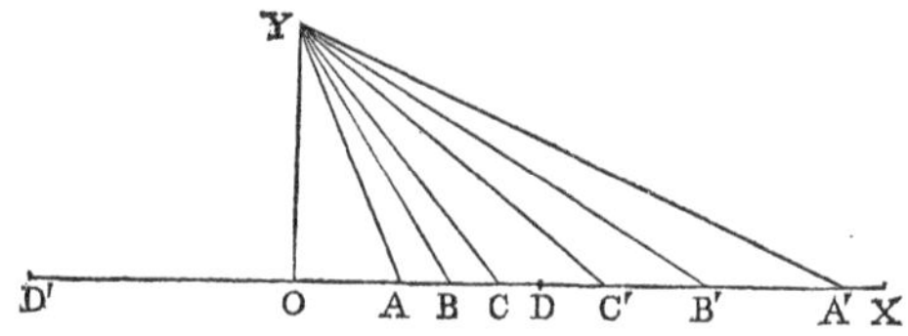

= A′YB′: similarly the ∠ BYC = B′YC′, &c.; ∴ the ∠s of the pencil (Y . ABCC′) = the ∠s of the pencil

(Y . A′B′C′C); and hence the anharmonic ratio of (Y . ABCC′) equal the anharmonic ratio of the pencil (Y . A′B′C′C).

Cor. 1.—If the point A moves towards O, the point A′ will move towards infinity.

Cor. 2.—The foregoing Demonstration will hold if some of the pairs of conjugate points be on the production of OX in the negative direction; that is, to the left of OY, while others are to the right, or in the positive direction.

Cor. 3.—If the points A, B, C, &c., be on one side of O, say to the right, their corresponding points A′, B′, C′, &c., may lie on the other side; that is, to the left. In this case the △s AYA′, BYB′, CYC′, &c., are all right-angled at Y; and the general Proposition holds for this case also, namely, *The anharmonic ratio of any four points is equal to the anharmonic ratio of their four conjugates.*

Cor. 4.—The anharmonic ratio of any four collinear points is equal to the anharmonic ratio of the four points which are inverse to them, with respect to any circle whose centre is in the line of collinearity.

Def.—*When two systems of three points each, such as* A, B, C; A′, B′, C′, *are collinear, and are so related that the anharmonic ratio of any four, which are not two couples of conjugate points, is equal to the anharmonic ratio of their four conjugates, the six points are said to be in* involution. *The point* O *conjugate to the point at infinity is called the* centre *of the involution. Again, if we take two points* D, D′, *one at each side of* O, *such that* $OD^2 = OD'^2 = k^2$, *it is evident that each of these points is its own conjugate. Hence they have been called, by* Townsend *and* Chasles, the double points *of the involution.* From these Definitions the following Propositions are evident:—

(1). *Any pair of homologous points, such as* A, A′, *are harmonic conjugates to the double points* D, D′.

(2). *Three pairs of points which have a common pair of harmonic conjugates form a system in involution.*

(3). *The two double points, and any two pairs of conjugate points, form a system in involution.*

(4). *Any line cutting three coaxal circles is cut in involution.*

Def.—*If a system of points in involution be joined to any point* P *not on the line of collinearity of the points, the six joining lines will have the anharmonic ratio of the pencil formed by any four rays equal to the anharmonic ratio of the pencil formed by their four conjugate rays. Such a pencil is called a* pencil in involution. *The rays passing through the double points are called the* double rays of the involution.

Prop. 7.—*If four points be collinear, they belong to three systems in involution.*

Dem.—Let the four points be A, B, C, D; upon AB and CD, as diameters, describe circles; then any circle coaxal with these will intersect the line of collinearity of A, B, C, D in a pair of points, which form an involution with the pairs A, B, C, D. Again, describe circles on the segments AD, BC, and circles coaxal with them will give us a second involution. Lastly, the circles described on CA, BD will give us a third system. The central points of these systems will be the points where the radical axes of the coaxal systems intersect the line of collinearity of the points.

Prop. 8.—The following examples will illustrate the theory of involution:—

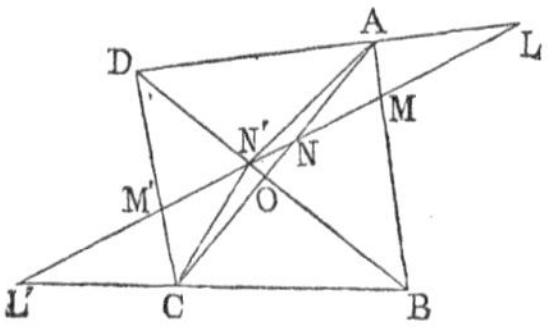

(1). *Any right line cutting the sides and diagonals of a quadrilateral is cut in involution.*

Dem.—Let ABCD be the quadrilateral, LL′ the transversal intersecting the diagonals in the points N, N′. Join AN′, CN′; then the anharmonic ratio of the pencil (A . LMNN′) = (A . DBON′) = (C . DBON′) = (C . M′L′ON) = (C . L′M′N′N).

(2). *A right line, cutting a circle and the sides of an inscribed quadrilateral, is cut in involution.*

Dem.—Join AR, AR′, CR, CR′; then the anharmonic ratio of the pencil (A . LRMR′) = (A . DRBR′) = (C . DRBR′) = (C . M′RL′R′) = (C . L′R′M′R).

Cor.—The points N, N′ belong to the involution.

(3). *If three chords of a circle be concurrent, their six points of intersection with the circle are in involution.*

Let AA′, BB′, CC′ be the three chords intersecting in the point O. Join AC, AC′, AB′, CB′; then the anharmonic ratio (A . CA′B′C′) = (B′ . CBAC′) = (B′ . C′ABC).

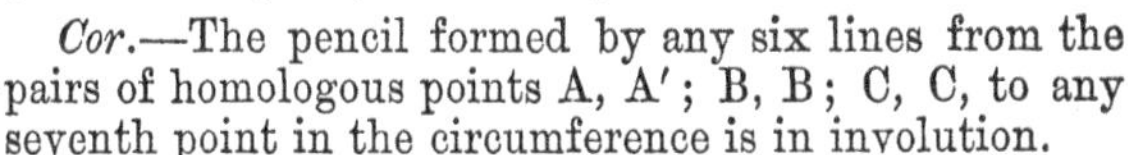

Cor.—The pencil formed by any six lines from the pairs of homologous points A, A′; B, B; C, C, to any seventh point in the circumference is in involution.

Prop. 9.—*If* O, O′ *be two fixed points on two given lines* OX, O′X′, *and if on* OX *we take any system of points* A, B, C, &c., *and on* O′X′ *a corresponding system* A′, B′, C′, &c., *such that the rectangles* OA . O′A′ = OB . O′B′ = OC . O′C′, &c., *equal constant, say* k^2; *then the anharmonic ratio of any four points on* OX *equal the anharmonic ratio of their four corresponding points on* O′X′.

This is evident by superposition of O′X′ on OX, so that the point O′ will coincide with O (see Prop. 7); then the two ranges on OX will form a system in involution.

Def.—*Two systems of points on two lines, such that the anharmonic ratio of any four points on one line is equal to the anharmonic ratio of their four corresponding points on the other, are said to be* homographic, *and the lines are said to be* homographically *divided. The points* O, O′ *are called the* centres of the systems.

Cor. 1.—The point O on OX is the point corresponding to infinity on O′X′; and the point O′ on O′X′ corresponds to infinity on OX.

Def.—*If the line* O′X′ *be superimposed on* OX, *but so that the point* O′ *will not coincide with* O, *the two systems of points on* OX *divide it homographically, and the points of one system which coincide with their homologous points of the other are called the* double points *of the homographic system.*

Prop. 10.—*Given three pairs of corresponding points of a line divided homographically, to find the double points.*

O A B C A′ B′ C′ X

Let A, A′; B, B′; C, C′, be the three pairs of corresponding points, and O one of the required double points; then the conditions of the question give us the anharmonic ratio

$$(OABC) = (O\,A'B'C');$$

therefore
$$\frac{OA \,.\, BC}{OB \,.\, AC} = \frac{OA' \,.\, B'C'}{O\,B' \,.\, A'C'}.$$

Hence
$$\frac{OA \,.\, OB'}{OA' \,.\, OB} = \frac{B'C' \,.\, AC}{BC \,.\, A'C'}$$

equal constant, say k^2.

Now OA . OB′, OA′ . OB are the squares of tangents drawn from O to the circles described on the lines AB′ and A′B as diameters; hence the ratio of these tangents is given; but if the ratio of tangents from a variable point to two fixed circles be given, the locus of the point is a circle coaxal with the given circles. Hence the point O is given as one of the points of intersection of a fixed circle with OX, and these intersections are the two double points of the homographic system.

If the three pairs of points be on a circle, the points of intersection of Pascal's line with the circle will be the double points required. For (see fig., Prop. 5), let D, B, F; A, E, C, be the two triads of points, and let the Pascal's line intersect the circle in the points P and Q; then it is evident that the pencil (A . PDBF) = (D . PAEC).

Cor.—If we invert the circle into a line, or *vice versâ*, the solution of either of the Problems we have given here will give the solution of the other.

Prop. 11.—*We shall conclude this Section with the solution of a few Problems by means of the double points of homographic division.*

(1). *Being given two right lines* L, L′, *it is required to place between them a line* AA′, *which will subtend given angles* Ω, Ω′ *at two given points* P, P′.

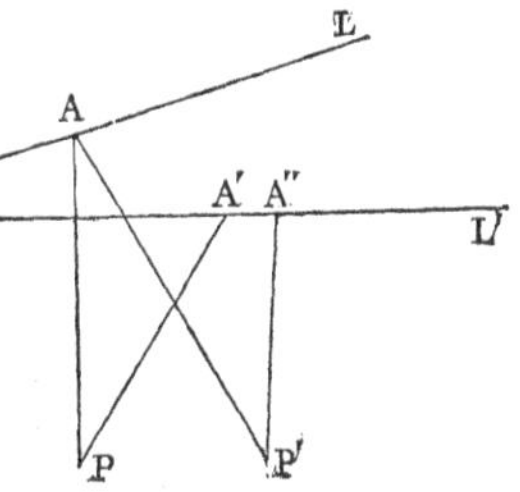

Solution.—Let us take arbitrarily any point A on L. Join PA, P′A, and make the ∠s APA′, APA″, respectively equal to the two given ∠s Ω, Ω′; then, when the point A moves along the line L, the points A′, A″ will form two homographic divisions on the line L′. The two double points of these divisions will give two solutions of the required Problem.

(2). *Being given a polygon of any number of sides, and as many points taken arbitrarily, it is required to inscribe in the polygon another polygon whose sides will pass through the given points.*

We shall solve this problem for the special case of a triangle; but it will be seen that the solution is perfectly general.

Let ABC be the given triangle; P, Q, R the given points. Take on AC any arbitrary point D. Join PD, intersecting AB in E; then join EQ, intersecting BC in F; lastly, join FR, intersecting AC in D′; then the two points D, D′ will evidently form two homographic divisions on AC, the two double points of which will be vertices of two triangles satisfying the question.

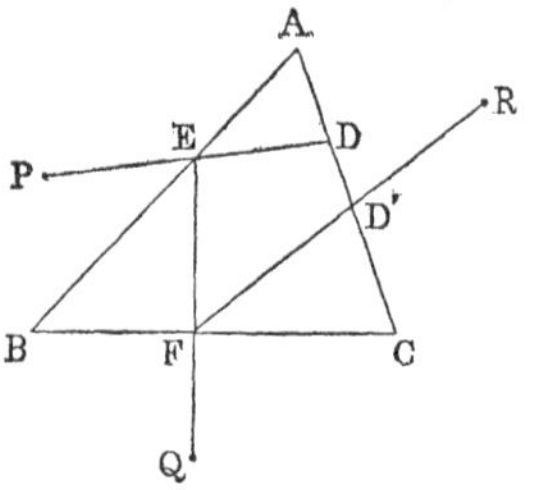

(3). *Being given three points* P, Q, R, *and two lines* L, L′, *it is required to describe a triangle* ABC *having* C *equal to a given angle, the vertices* A *and* B *on the given lines* L, L′, *and the sides passing through the given points.*

Solution.—Through the point R draw any line meeting the two lines L, L′ in the points a, b. Join Pb, and from Q draw Qa' making the required angle C with Pb; the two points a, a' will form two homographic divisions on L, the double points of which will give two solutions of the required question.

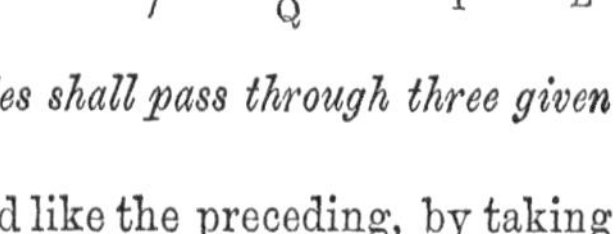

(4). *To inscribe in a circle a triangle whose sides shall pass through three given points.*

This is evidently solved like the preceding, by taking three false positions, and finding the double points of the two homographic systems of points.

(5). *The problem of describing a circle touching three given circles can be solved at once by the method of taking three false positions, and finding double points, as follows :—*

Let A, B, C be the centres of the three given circles; Z the required tangential circle ; α, β, γ the points of contact : then the triangles ABC, $\alpha\beta\gamma$ are in perspective, the centre of Z being their centre of perspective, and the axis of similitude of the three given circles being their axis of perspective. Let A′, B′, C′ be the three centres

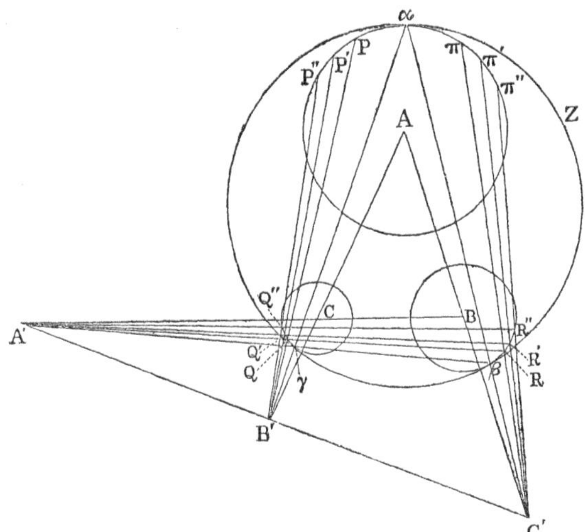

of similitude. Then take any three points P, P′, P″ in the circle A; join them to the point B′, cutting the circle C in the points Q, Q′, Q″: again, join these points to A′, and let the joining lines cut the circle B in the points R, R′, R″ : lastly, join R, R′, R″ to C′, cutting the circle A in the points π, π', π'' ; then α will be such that the anharmonic ratio (αPP′P″) will be equal to the anharmonic ratio ($\alpha\,\pi\,\pi'\pi''$). Hence the problem is solved.

Exercises.

1. The eight circles which touch three given circles may be divided into two tetrads—say X, Y, Z, W ; X′, Y′, Z′, W′—of which one is the inverse of the other with respect to the circle cutting the three given circles orthogonally.

2. Any two circles of the first tetrad, and the two corresponding circles of the second, have a common tangential circle.

3. Any three circles of either tetrad, and the non-corresponding circle of the other tetrad, have a common tangential circle.

4. Prove by means of the extension of Ptolemy's Theorem (the middle points of the sides being regarded as very small circles) that these point-circles, and the inscribed circle, or any of the escribed circles, have a common tangential circle.

5. The anharmonic ratios of the four points of contact of the "nine-points circle" with the inscribed and the escribed circles are respectively

$$\frac{a^2 - b^2}{a^2 - c^2}, \quad \frac{b^2 - c^2}{b^2 - a^2}, \quad \frac{c^2 - a^2}{c^2 - b^2}.$$

SECTION VII.

Theory of Poles and Polars, and Reciprocation.

Prop. 1.—*If four points be collinear, their anharmonic ratio is equal to the anharmonic ratio of their four polars.*

This Proposition may be proved exactly the same as Proposition 8, Section III. Thus (see fig., Prop. 8. Section III.) the pencil (O . A′B′C′D′) = (P . A′B′C′D′); but the pencil (O . A′B′C′D′) = anharmonic ratio of the four points A, B, C, D, and the pencil (P . A′B′C′D′) consists of the four polars. Hence the Proposition is proved.

The two following Propositions are interesting applications of this Proposition:—

(1). *If two triangles be self-conjugate with respect to a circle, any two sides are divided equianharmonically by the four remaining sides; and any two vertices are subtended equianharmonically by the four remaining vertices.*

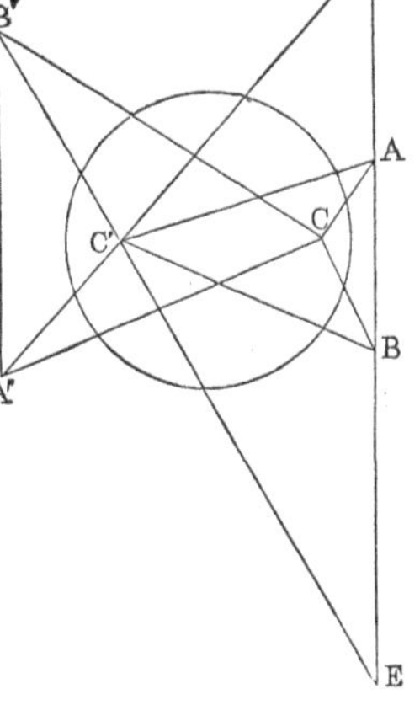

Let ABC, A′B′C′ be the two self-conjugate △s; it is required to prove that the pencil (C . ABA′B′) = (C′ . ABA′B′).

Dem.—Let A′C′, B′C′ meet AB produced in D and E. Join A′C, B′C, AC′, BC′. Now, since A′C′ is the polar of B′, and AB the polar of C, their point of intersection D is the pole of B′C (see *Cor.*, Prop. 25, Section I., Book III.). In like manner, the point E is the pole of A′C; hence the four points B, A, E, D are the poles of the four lines CA, CB, CA′, CB′. Therefore the anharmonic ratio of the four points B,

A, E, D is equal to the anharmonic ratio of the pencil (C . ABA′B′). Again, the points B, A, E, D are the intersections of the line AB with the pencil (C . ABA′B′); therefore the pencil (C . ABA′B′) = (C′ . ABA′B′).

We have proved the second part of our Proposition, and the first follows from it by the theorem of this Article.

(2). *If two triangles be such that the sides of one are the polars of the vertices of the other, they are in perspective.*

Dem.—Let the three sides of the △ ABC be the polars of the corresponding vertices of the △ A′B′C′, and let the corresponding sides meet in the points X, Y, Z respectively. Now, since AB is the polar of C′, and B′C′ the polar of A, the point D is the pole of AC′ (*Cor.*, Prop. 25, Section I., Book III.). In like manner the point X is the pole of AA′, and the points B′, C′ are, by hypothesis, the poles of the lines AC, AB. Hence the anharmonic ratio of the points B′, C′, D, X = the pencil (A . YZC′A′) = the pencil (Z . YAC′A′). Again, the anharmonic ratio B′C′DX = the pencil (Z . A′C′AX) = (Z . XAC′A′). Hence (Z . YAC′A′) = (Z . XAC′A′); ∴ the lines XZ, ZY form one right

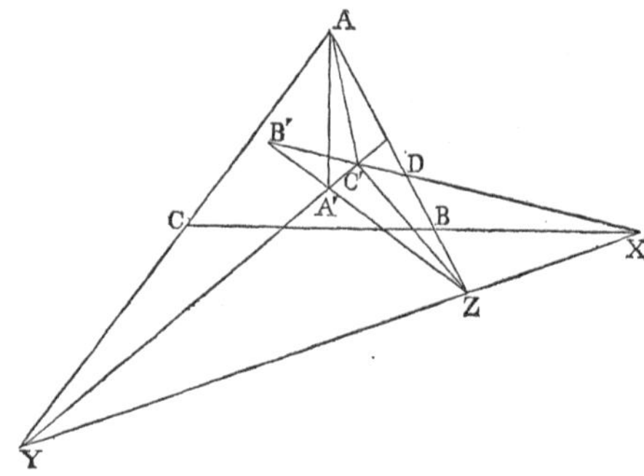

line; therefore the intersection of corresponding sides of the triangles are collinear. Hence they are in perspective.

Prop. 2.—*If two variable points* A, A′, *one on each of two lines given in position, subtend an angle of constant magnitude at a given point* O, *the locus of the pole of the line* AA′ *with respect to a given circle* X, *whose centre is* O, *is a circle.*

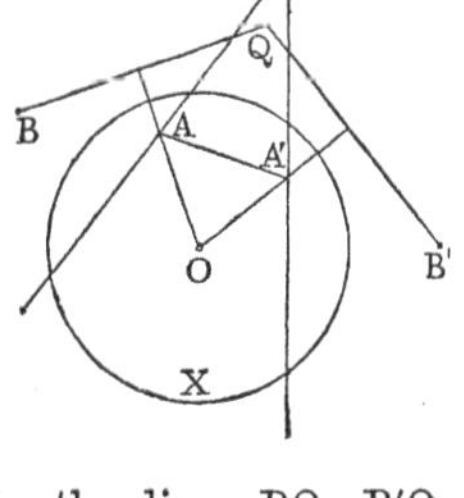

Dem.—Let AI, A′I be the lines given in position, and let B, B′, Q be the poles of the three lines AI, A′I, and AA′ with respect to X; then the points B, B′ are fixed, and the lines BQ, B′Q are the polars of the points A, A′; ∴ the lines OA, OA′ are respectively ⊥ to the lines BQ, B′Q; hence the ∠ BQB′ is the supplement of the ∠ AOA′; therefore BQB′ is a given angle, and the points B, B′ are fixed; therefore the locus of the point Q is a circle.

Prop. 3.—*For two homographic systems of points on two lines given in position there exist two points, at each of which the several pairs of corresponding points subtend equal angles.*

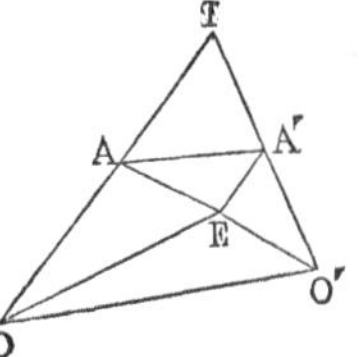

Dem.—Let A, A′ be two corresponding points on the lines AI, A′I; and let O, O′ be the points on the lines AI, A′I which correspond to the points at infinity on A′I, AI respectively; then (see Prop. 10, Section IV., Book VI.) the rectangle OA . O′A′ = constant, say k^2. Join OO′, and describe the triangle OEO′ (see Prop. 15, Section I., Book VI.) having the rectangle OE . O′E of its sides = k^2, and having the difference of its base ∠s equal difference of base ∠s of the △ OIO′. Then E, the vertex of this △, will be one of the points required. For it is evident from the construction that OE . O′E = OA . O′A, and that the ∠ AOE = A′O′E;

∴ the △s AOE, A′O′E are equiangular; ∴ the ∠ OAE = A′EO′; ∴ if the points A, A′ change position, the lines EA, EA′ will revolve in the same direction, and through equal angles. Hence the ∠ AEA′ is constant.

In the same manner, another point F can be found on the other side of OO′ such that the ∠ AFA′ is constant.

Cor. 1.—Since the line AA′ subtends a constant angle at E, the locus of the pole of AA′ with respect to a circle whose centre is E is a circle. Hence the properties of lines joining corresponding points on two lines divided homographically may be inferred from the properties of a system of points on a circle.

Cor. 2.—Since when A′ goes to infinity A coincides with O, then OA is one of the lines joining corresponding points. And so in like manner is O′A′, and the poles of these lines will be points on the circle which is the locus of the pole of AA′.

Cor. 3.—The locus of the foot of the perpendicular from E on the line AA′ is a circle, namely, the inverse of the circle which is the locus of the pole of AA′.

Cor. 4.—If two lines be divided homographically, any four lines joining corresponding points are divided equianharmonically by all the remaining lines joining corresponding points. This follows from the fact that any four points on a circle are subtended equianharmonically by all the remaining points of the circle.

Prop. 4.—*If any figure* A *be given, by taking the pole of every line, and the polar of every point in it with respect to any arbitrary circle* X, *we can construct a new figure* B, *which is called the reciprocal of* A *with respect to* X. *Thus we see that to any system of collinear points or concurrent lines of* A *there will correspond a system of concurrent lines or collinear points of* B; *and to any pair of lines divided homographically in* A *there will correspond in* B *two homographic pencils of lines. Lastly, the angle which any two points of* A *subtend at the centre of the reciprocating circle is equal to the angle made by their polars in* B.

Hence it is evident that, from theorems which hold for A, we can get other theorems which are true for B. This method, which is called reciprocation, is due to Poncelet, and is one of the most important known to Geometers.

We give a few Theorems proved by this method:—

(1). *Any two fixed tangents to a circle are cut homographically by any variable tangent.*

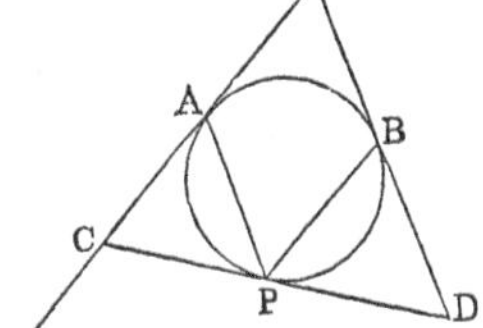

Dem.—Let AT, BT be the two fixed tangents touching the circle at the fixed points A and B, and CD a variable tangent touching at P. Join AP, BP. Now AP is the polar of C, and BP the polar of D; and if the point P take four different positions, the point C will take four different positions, and so will the point D; and the anharmonic ratio of the four positions of C equal the anharmonic ratio of the pencil from A to the four positions of P (Prop. 1). Similarly the anharmonic ratio of the four positions of D equal the anharmonic ratio of the pencil from B to the four positions of P; but the pencil from A equal the pencil from B; therefore the anharmonic ratio of the four positions of C equal the anharmonic ratio of the four positions of D.

(2). *Any four fixed tangents to a circle are cut by any fifth variable tangent in four points whose anharmonic ratio is constant.*

Dem.—The lines joining the point of contact of the variable tangent to the points of contact of the fixed tangents are the polars of the points of intersection of the variable tangent with the fixed tangents; but the anharmonic ratio of the pencil of four lines from a variable point to four fixed points on a circle is constant; hence the anharmonic ratio of their four poles—that is, of the four points in which the variable tangent cuts the fixed tangent—is constant.

(3). *Lines drawn from any variable point in the plane of a quadrilateral to the six points of intersection of its four sides form a pencil in involution.*

This Proposition is evidently the reciprocal of (1), Prop. 9, Section VI. The following is a direct proof: Let ABCD be the quadrilateral, and let its opposite sides meet in the points E and F, and let O be the point in the plane of the quadrilateral; the pencil from O to the points A, B, C, D, E, F is in involution.

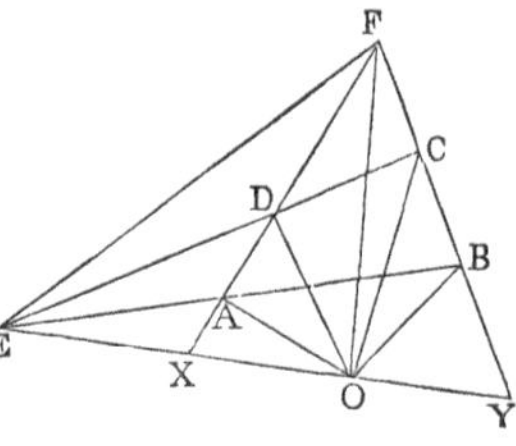

Dem.—Join OE, cutting the sides AD, BC in X and Y. Join EF. Now, the pencil (O . XADF) = (E . XADF) = (E . YBCF) = (O . YBCF) = (O . XBCF); ∴ (O . EADF) = (O . EBCF). Hence the pencil is in involution.

(4). *If two vertices of a triangle move on fixed lines, while the three sides pass through three collinear points, the locus of the third vertex is a right line.* Hence, reciprocally, *If two sides of a triangle pass through fixed points, while the vertices move on three concurrent lines, the third side will pass through a fixed point.*

(5). *To describe a triangle about a circle, so that its three vertices may be on three given lines.* This is solved by inscribing in the circle a triangle whose three sides shall pass through the poles of the three given lines, and drawing tangents at the angular points of the inscribed triangle.

(4). Brianchon's Theorem.—*If a hexagon be described about a circle, the three lines joining the opposite angular points are concurrent.*

This is the reciprocal of Pascal's Theorem: we prove it as follows:—

Let ABCDEF be the circumscribed hexagon; the three diagonals AD, BE, CF are concurrent. For, let the points of contact be G, H, I, J, K, L. Now since A is the pole of GH, and D the pole of JK, the line AD is the polar of the point of intersection of the opposite sides GH and JK of the inscribed hexagon. In like manner, BE is the polar of the point of intersection of the lines HI, KL, and CF the polar of the point of intersection of IJ and LG; but the intersections of the three pairs of opposite sides of the inscribed hexagon, viz., GH, JK; HI, KL; IJ, LG, are, by Pascal's Theorem, collinear; therefore their three polars AD, BE, CF, are concurrent.

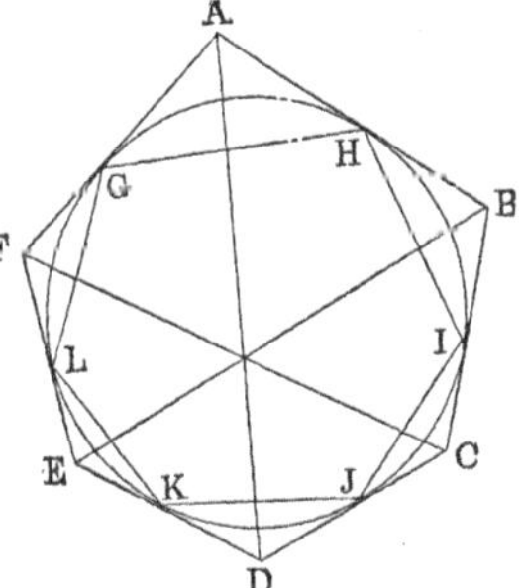

(7). *If two lines be divided homographically, two lines joining homologous points can be drawn, each of which passes through a given point.*

For, if AA′ (see fig., Prop. 3) pass through a given point P, join EP, and let fall a ⊥ EG on AA′; then (*Cor*. 2, Prop. 3) the locus of the point G is a ⊙; and since EGP is a right angle, the ⊙ described on EP as diameter passes through G; hence G is the point of intersection of two given ⊙s; and since two ⊙s intersect in two points, we see that two lines joining homographic points can be formed, each passing through P. Now, if we reciprocate the whole diagram with respect to a circle whose centre is P, the reciprocals of the points A, A′ will be parallel lines. Hence we have the following theorem in a system of two homographic pencils of rays:—*There exist two pairs of homologous rays which are parallel to each other.*

Cor.—There are two directions in which transversals can be drawn, intersecting two homographic pencils of rays so as to be divided proportionally, namely, parallel to the pairs of homologous rays which are parallel.

(8). If we reciprocate Prop. 3 we have the following theorem:—*Being given a fixed point, namely, the centre of the circle of reciprocation and two homographic pencils of rays, two lines can be found (the polars of the points* E *and* F *in* Prop. 3), *so that the portions intercepted on each by homologous rays of the pencils will subtend an angle of constant magnitude at the given point.*

SECTION VIII.

Miscellaneous Exercises.

1. The lines from the angles of a △ to the points of contact of any ⊙ touching the three sides are concurrent.

2. Three lines being given in position, to find a point in one of them, such that the sum of two lines drawn from it, making given angles with the other two, may be given.

3. From a given point in the diameter of a semicircle produced to draw a line cutting the semicircle, so that the lines may have a given ratio which join the points of intersection to the extremities of the diameter.

4. The internal and external bisectors of the vertical angle of a △ meet the base in points which are harmonic conjugates to the extremities.

5. The rectangle contained by the sides of a △ is greater than the square of the internal bisector of the vertical angle by the rectangle contained by the segments of the base.

6. State the corresponding theorem for the external bisector.

7. Given the base and the vertical angle of a △, find the following loci:—

(1). Of the intersection of perpendiculars.

(2). Of the centre of any circle touching the three sides.

(3). Of the intersection of bisectors of sides.

8. If a variable ⊙ touch two fixed ⊙s, the tangents drawn to it from the limiting points have a constant ratio.

9. The ⊥ from the right angle on the hypotenuse of a right-angled Δ is a harmonic mean between the segments of the hypotenuse made by the point of contact of the inscribed circle.

10. If a line be cut harmonically by two ⊙s, the locus of the foot of the ⊥, let fall on it from either centre, is a ⊙, and it cuts any two positions of itself homographically (see Prop. 3, *Cor.* 2, Section VII.).

11. Through a given point to draw a line, cutting the sides of a given Δ in three points, such that the anharmonic ratio of the system, consisting of the given point and the points of section, may be given.

12. If squares be described on the sides of a Δ and their centres joined, the area of the Δ so formed exceeds the area of the given triangle by ⅛th part of the sum of the squares.

13. The locus of the centre of a ⊙ bisecting the circumferences of two fixed ⊙s is a right line.

14. Divide a given semicircle into two parts by a ⊥ to the diameter, so that the diameters of the ⊙s described in them may be in a given ratio.

15. The side of the square inscribed in a Δ is half the harmonic mean between the base and perpendicular.

16. The ⊙s described on the three diagonals of a quadrilatera are coaxal.

17. If X, X′ be the points where the bisectors of the ∠ A of a Δ and of its supplement meet the side BC, and if Y, Y′; Z, Z′, be points similarly determined on the sides CA, AB; then

$$\frac{1}{XX'} + \frac{1}{YY'} + \frac{1}{ZZ'} = 0;$$

and

$$\frac{a^2}{XX'} + \frac{b^2}{YY'} + \frac{c^2}{ZZ'} = 0.$$

18. Prove Ptolemy's Theorem, and its converse, by inversion

19. A line of given length slides between two fixed lines: find the locus of the intersection of the ⊥s to the fixed lines from the extremities of the sliding line, and of the ⊥s on the fixed lines from the extremities of the sliding line.

20. If from a variable point P ⊥s be drawn to three sides of a Δ; then, if the area of the Δ formed by joining the feet of these ⊥s be given, the locus of P is a circle.

21. If a variable ⊙ touch two fixed ⊙s, its radius varies as the square of the tangent drawn to it from either limiting point.

22. If two ⊙s, whose centres are O, O′, intersect, as in Euclid (I. 1), and OO′ be joined, and produced to A, and a ⊙ GDH be described, touching the ⊙s whose centres are O, O′, and also touching the line AO; then, if we draw the radical axis EE′ of the ⊙s, intersecting OO′ in C, and the diameter DF of the ⊙ GHD, and join EF, the figure CDFE is a square.

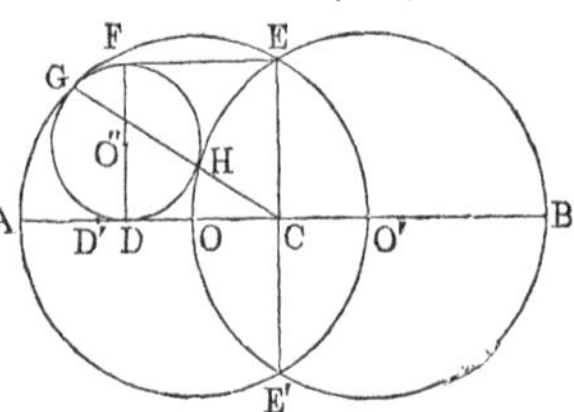

Dem.—The line joining the points of contact G and H will pass through C, the internal centre of similitude of the ⊙s O, O′; therefore $CG \cdot CH = CE^2$; but $CD^2 = CG \cdot CH$; therefore CD = CE.

Again, let O″ be the centre of GDH, and D′ the middle point of AO; then the ⊙ whose centre is D′ and radius D′A touches the ⊙s O, O′; hence (by Theorem 7, Section V.) the ⊥ from O″ on EE′ : O″D : : CD′ : D′A; that is, in the ratio of 2 : 1. Hence the Proposition is proved.

23. If a quadrilateral be circumscribed to a ⊙, the centre and the middle points of the diagonals are collinear.

24. If one diagonal of a quadrilateral inscribed in a ⊙ be bisected by the other, the square of the latter = half the sum of the squares of the sides.

25. If a △ given in species moves with its vertices on three fixed lines, it marks off proportional parts on these lines.

26. Through the point of intersection of two ⊙s draw a line so that the sum or the difference of the squares of the chords of the ⊙s shall be given.

27. If two ⊙s touch at A, and BC be any chord of one touching the other; then the sum or difference of the chords AB, AC bears to the chord BC a constant ratio. Distinguish the two cases.

28. If ABC be a △ inscribed in a ⊙, and if a ∥ to AC through the pole of AB meet BC in D, then AD is = CD.

29. The centres of the four ⊙s circumscribed about the △s formed by four right lines are concyclic.

30. Through a given point draw two transversals which shall intercept given lengths on two given lines.

31. If a variable line meet four fixed lines in points whose anharmonic ratio is constant, it cuts these four lines homographically.

32. Given the ⊥ CD to the diameter AB of a semicircle, it is required to draw through A a chord, cutting CD in E and the semicircle in F, such that the ratio of CE : EF may be given.

33. Draw in the last construction the line AEF so that the quadrilateral CEFB may be a maximum.

34. The ⊙ described through the centres of the three escribed ⊙s of a plane △, and the circumscribed ⊙ of the same △, will have the centre of the inscribed ⊙ of the △ for one of their centres of similitude.

35. The ⊙s on the diagonals of a complete quadrilateral cut orthogonally the ⊙ described about the △ formed by the three diagonals.

36. When the three ⊥s from the vertices of one △ on the sides of another are concurrent, the three corresponding ⊥s from the vertices of the latter, on the sides of the former, are concurrent.

37. If a ⊙ be inscribed in a quadrant of a ⊙; and a second ⊙ be described touching the ⊙, the quadrant, and radius of quadrant; and a ⊥ be let fall from the centre of the second ⊙ on the line passing through the centres of the first ⊙ and of the quadrant; then the △ whose angular points are the foot of the ⊥, the centre of the quadrant, and the centre of the second ⊙, has its sides in arithmetical progression.

38. In the last Proposition, the ⊥s let fall from the centre of the second ⊙ on the radii of the quadrants are in the ratio of 1 : 7.

39. When three ⊙s of a coaxal system touch the three sides of a △ at three points, which are either collinear or concurrently connectant with the opposite vertices, their three centres form, with those of the three ⊙s of the system which pass through the vertices of the △, a system of six points in involution.

40. If two ⊙s be so placed that a quadrilateral may be inscribed in one and circumscribed to the other, the diagonals of the quadrilateral intersect in one of the limiting points.

41. If from a fixed point ⊥s be let fall on two conjugate rays of a pencil in involution, the feet of the ⊥s are collinear with a fixed point.

42. Miquel's Theorem.—If the five sides of any pentagon ABCDE be produced, forming five △s external to the pentagon,

the ⊙s described about these △s intersect in five points A″, B″, C″, D″, E″, which are concyclic.

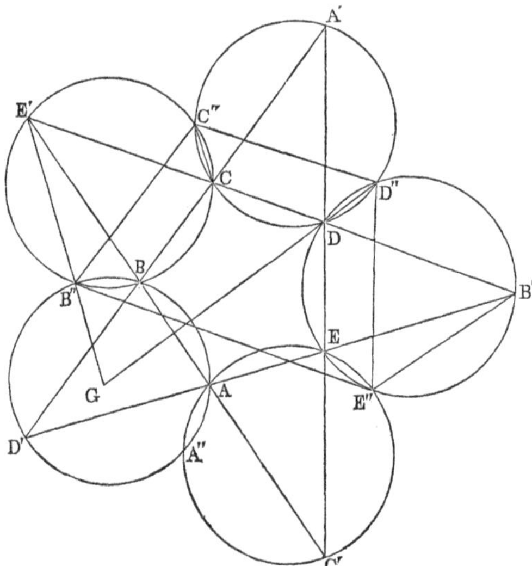

Dem.—Join E″B′, E″D″, D″C″, C″B″, C″C; join also D″D and E′B″, and let them produced meet in G. Now, consider the △ AB′E′, it is evident the ⊙ described about it (*Cor.* 3, Prop. 12, Book III.) will pass through the points E″, B″; hence the four points E″, B′, E′, B″ are concyclic; ∴ the ∠ GB″E″ = E′B′E″; but E′B′E″ = GD″E″; ∴ ∠ GB″E″ = GD″E″. Hence the ⊙ through the points B″, D″, E″ passes through G.

Again, since the figure CDD″C″ is a quadrilateral in a ⊙, the ∠ GDE′ = D″C″C, and the ∠ GE′D = B″C″C (III. 21); ∴ ∠ B″C″D″ = GDE′ + GE′D. To each add ∠ E′GD, and we see that the figure GD″C″B″ is a quadrilateral in a ⊙; hence the ⊙ through the points B″, D″, E″ passes through C″. In like manner it passes through A″. Hence the five points A″, B″, C″, D″, E″ are concyclic.

43. If the product of the tangents, from a variable point P to two given ⊙s, has a given ratio to the square of the tangent from P to a third given ⊙ coaxal with the former, the locus of P is a circle of the same system.

44. Through the vertices of any △ are drawn any three parallel lines, and through each vertex a line is drawn, making the same ∠ with one of the adjacent sides which the parallel makes with the other; these three lines are concurrent. Required the locus of the point in which they meet.

45. If from any point in a given line two tangents be drawn to a given ⊙, X, and if a ⊙, Y, be described touching X and the two tangents, the envelope of the polar of the centre of Y with respect to X is a circle.

46. The extremities of a variable chord XY of a given ⊙ are joined to the extremities of a fixed chord AB; then, if m AX . AY + n BX . BY be given, the envelope of XY is a circle.

47. If A, A′ be conjugate points of a system in involution, and if AQ, A′Q be ⊥ to the lines joining A, A′ to any fixed point P, it is required to find the locus of Q.

48. If a, a', b, b', c, c', be three pairs of conjugate points of a system in involution; then,

(1). $$ab' \,.\, bc' \,.\, ca' = - a'b \,.\, b'c \,.\, c'a.$$

(2). $$ab' \,.\, bc \,.\, c'a' = - a'b \,.\, b'c' \,.\, ca.$$

(3). $$\frac{ab \,.\, ab'}{ac \,.\, ac'} = \frac{a'b \,.\, a'b'}{a'c \,.\, a'c'}.$$

49. Construct a right-angled △, being given the sum of the base and hypotenuse, and the sum of the base and perpendicular.

50. Given the perimeter of a right-angled △ whose sides are in arithmetical progression: construct it.

51. Given a point in the side of a △; inscribe in it another △ similar to a given △, and having one ∠ at the given point.

52. Given a point D in the base AB produced of a given △ ABC; draw a line EF through D cutting the sides so that the area of the △ EFC may be given.

53. Construct a △ whose three ∠s shall be on given ⊙s, and whose sides shall pass through three of their centres of similitude.

54. From a given point O three lines OA, OB, OC are drawn to a given line ABC; prove that if the radii of the ⊙s inscribed in OAB, OBC are given, the radius of the ⊙ inscribed in OAC will be determined.

55. Equal portions OA, OB are taken on the sides of a given right ∠ AOB, the point A is joined to a fixed point C, and a ⊥ let fall on AC from B: the locus of the foot of this ⊥ is a circle.

56. If a segment AB of a given line be cut in a given anharmonic ratio in two variable points X, X′, then the anharmonic ratio of any four positions of X will be equal to the anharmonic ratio of the four corresponding positions of X′.

57. If a variable △ inscribed in a ⊙, X, whose radius is R, has two of its sides touching another ⊙, Y, whose radius is r, and whose centre is distant from the centre of X by δ; then the distance of the centre of the ⊙ coaxal with X and Y, which is the envelope of the third side of the △ from the centre of X,

$$= r^2\,\delta \div \frac{(R^2 - \delta^2)^2}{4R^2}.$$

58. In the same case the radius of the ⊙ which is the envelope of the third side is

$$\frac{r^2\,(R - \rho) - R\rho^2}{\rho^2};$$

where $$\rho = \frac{R^2 - \delta^2}{2R}.$$

59. If two tangents be drawn to a ⊙, the points where any third tangent is cut by these will be harmonic conjugates to the point of contact and the point where it is cut by the chord of contact.

60. If two points be inverse to each other with respect to any ⊙, then the inverses of these will be inverse to each other with respect to the inverse of the ⊙. Hence it follows that if two figures be inverse to each other with respect to any ⊙, their inverses will be inverse to each other with respect to the inverse of the circle.

61. MALFATTI'S PROBLEM.—To inscribe in a △ three ⊙s which touch each other, and each of which touches two sides of the △.

Analysis.—Let L, M, N be the points of contact of three ⊙s which touch one another, and each touch two sides of the △ ABC; draw the common tangents DE, FG, HI to these ⊙s at their points of contact L, M, N; then, since these lines are the radical axes of the ⊙s taken in pairs, they are concurrent: let them meet in K.

Now, it is evident that FH − HD = FO − DP = FM − DL = FK − DK. Hence H is the point of contact with FD of the ⊙ described in the △ FKD. In like manner, E and G are the points of contact of ⊙s which touch the triads of lines IK, KF, AC; and IK, DK, AB, respectively.

Again, HN = HP = QL, and NS = ER = EL; ∴ HS = EQ; ∴ (see 6, Prop. 1, Section V.) the tangents at E and H to the ⊙s

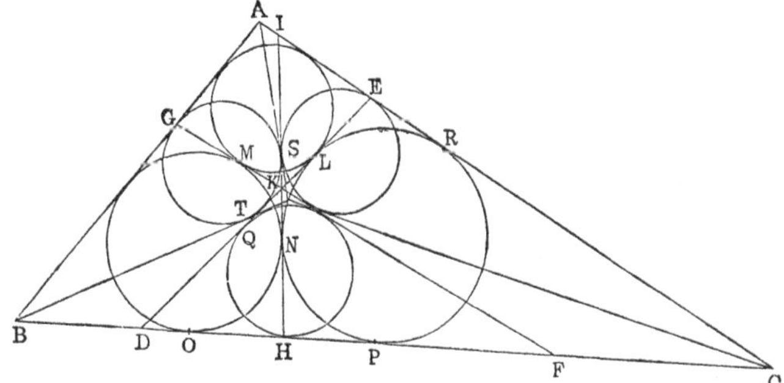

ES and HQ meet on their ⊙ of similitude; ∴ C is a point on the ⊙ of similitude of the ⊙s ES and HQ; and therefore these ⊙s subtend equal ∠ s at C. Also, three common tangents of the ⊙s HQ, ES, PNR, viz., QL, SN, KF, are concurrent; ∴ (see Ex. 48, Section II., Book III.) C must be the point of concurrence of three other common tangents to the same ⊙s. Hence the second transverse common tangent to HQ and ES must pass through C; and since C is a point on their ⊙ of similitude, this transverse common tangent must bisect the ∠ ACB. In like manner it is proved that the bisectors of A and B are transverse common tangents to the ⊙s ES and GT, and to HQ and GT, respectively. Hence, we have the following elegant construction:—Let V be the point of concurrence of the three bisectors of the ∠ s of the Δ ABC. In the Δ s VAB, VBC, VCA, describe three ⊙s: these ⊙s will evidently, taken in pairs, have VB, VC, VA as transverse common tangents; then to the same pairs of ⊙s draw the three other transverse tangents; these will be respectively ED, GF, HI; and the ⊙s described touching the triads of lines AB, AC, ED; AB, BC, GF; AB, BC, HI, will be the required circles.

This construction is due to Steiner, and the foregoing simple and elementary proof to Dr. Hart (see *Quarterly Journal*, vol. i. p. 219).

62. If a transversal passing through a fixed point O cut any number of fixed lines in the points A, B, C, &c., and if P be a point such that

$$\frac{1}{OP} = \frac{1}{OA} + \frac{1}{OB} + \frac{1}{OC} + \&c.,$$

the locus of P is a right line.

63. The sum of the squares of the radii of the four ⊙s, cutting orthogonally the inscribed and escribed ⊙s of a plane △, taken three by three, is equal to the square of the diameter of the circumscribed ⊙.

64. Describe through two given points a ⊙ cutting a given arc of a given ⊙ in a given anharmonic ratio.

65. All ⊙s which cut three fixed ⊙s at equal ∠s form a coaxal system.

66. Being given five points and a line, find a point on the line, so that the pencil formed by joining it to the five given points shall form an involution with the line itself.

67. If a quadrilateral be inscribed in a circle, the circle described on the third diagonal as diameter will be the circle of similitude of the circles described on the other diagonals as diameters.

68. If ABC be any △, B′C′ a line drawn ∥ to the base BC; then, if O, O′ be the escribed ⊙s to ABC, opposite the ∠s B and C respectively, O_1 the inscribed ⊙ of AB′C′, and O'_1 the escribed ⊙ opposite the ∠ A; then, besides the lines AB, AC, which are common tangents, O, O′, O_1, O'_1, are all touched by two other circles.

69. When two ⊙s intersect orthogonally, the locus of the point whence four tangents can be drawn to the ⊙s, and forming a harmonic pencil, consists of two lines, viz., the polars of the centre of similitude of the two circles.

70. If two lines be divided homographically in the two systems of points *a*, *b*, *c*, &c., *a′*, *b′*, *c′*, &c., then the locus of the points of intersection of *ab′*, *a′b*, *ac′*, *a′c*, *ad′*, *a′d*, &c., is a right line.

71. Being given two homographic pencils, if through the point of intersection of two corresponding rays we draw two transversals, which meet the two pencils in two series of points, the lines joining corresponding points of intersection are concurrent.

72. Inscribe a △ in a ⊙ having two sides passing through two given points, and the third ∥ to a given line.

73. If two △s be described about a ⊙, the six angular points are such that any four are subtended equianharmonically by the other two.

74. Given four points A, B, C, D on a given line, find two other points X, Y, so that the anharmonic ratios (ABXY), (CDXY) may be given.

75. If two quadrilaterals have the same diagonals, the eight points of intersection of their sides are such that any four are subtended equianharmonically by the other four.

76. Given three rays A, B, C, find three other rays X, Y, Z through the same vertex O, so that the anharmonic ratios of the pencils (O . ABXY), (O . BCYZ), (O . CAZX), may be given.

77. If a △ similar to that formed by the centres of three given ⊙s slide with its three vertices on their circumferences, the vertices divide the ⊙s homographically.

78. Find the locus of the centre of a ⊙, being given that the polar of a given point A passes through a given point B, and the polar of another given point C passes through a given point D.

79. If a △ be self-conjugate with respect to a given ⊙, the ⊙ described about the △ is orthogonal to another given circle.

80. The ⊙s self-conjugate to the △s formed by four lines are coaxal.

81. The pencil formed by lines ∥ to the sides and diagonals of a quadrilateral is involution.

82. If four ⊙s be co-orthogonal, that is, have a common orthogonal ⊙, their radical axes form a pencil in involution.

83. In a given ⊙ to inscribe a △ whose sides shall divide in a given anharmonic ratio given arcs of the circle.

84. When four ⊙s have a common point of intersection, their six radical axes form a pencil in involution.

85. The pencil formed by drawing tangents from any point in their radical axis to two ⊙ s, and drawing two lines to their centres of similitude, is in involution.

*86. If a pair of the opposite ∠ s of a quadrilateral be equal to a right ∠, then the sum of the squares of the rectangles contained by the opposite sides is equal to the square of the rectangle contained by the diagonals.

87. Prove that the problem 17, page 38, "To inscribe in a given △ DEF, a △ given in species whose area shall be a minimum," admits of two solutions; and also that the point O′ in the second solution, which corresponds to O in the first, is the inverse of O with respect to the circle which circumscribes the △ DEF.

88. The line joining the intersection of the ⊥s of a △ to the centre of a circumscribed ⊙ is ⊥ to the axis of perspective of the given △, and the △ formed by joining the feet of the ⊥s.

---

* This Theorem is due to Bellavitis. See his *Méthode des Equipollences.*

89. If two ⊙s whose radii are R, R′, and the distance of whose centres is δ, be such that a hexagon can be inscribed in one and circumscribed to the other; then

$$\frac{1}{(R^2-\delta^2)^2+4R'^2R\delta}+\frac{1}{(R^2-\delta^2)^2-4R'^2R\delta}$$
$$=\frac{1}{2R'^2(R^2+\delta^2)-(R^2-\delta^2)^2}.$$

90. In the same case, if an octagon be inscribed in one and circumscribed to the other,

$$\left\{\frac{1}{(R^2-\delta^2)^2+4R'^2R\delta}\right\}^2+\left\{\frac{1}{(R^2-\delta^2)^2-4R'^2R\delta}\right\}^2$$
$$=\left\{\frac{1}{2R'^2(R^2+\delta^2)-(R^2-\delta)^2}\right\}^2.$$

91. If a variable ⊙ touch two fixed ⊙s, the polar of its centre with respect to either of the fixed ⊙s touches a fixed circle.

92. If a ⊙ touch three ⊙s, the polar of its centre, with respect to any of the three ⊙s, is a common tangent to two circles.

*93. Prove that the Problem, to inscribe a quadrilateral, whose perimeter is a minimum in another quadrilateral, is indeterminate or impossible, according as the given quadrilateral has the sum of its opposite angles equal or not equal to two right angles.

94. If a quadrilateral be inscribed in a ⊙, the lines joining the feet of the ⊥s, let fall on its sides from the point of intersection of its diagonals, will form an inscribed quadrilateral Q of minimum perimeter; and an indefinite number of other quadrilaterals may be inscribed whose sides are respectively equal to the sides of Q, the perimeter of each of them being equal to the perimeter of Q.

95. The perimeter of Q is equal to the rectangle contained by the diagonals of the original quadrilateral divided by the radius of the circumscribed circle.

96. Being given four lines forming four △s, the sixteen centres of the inscribed and escribed ⊙s to these △s lie four by four on four coaxal circles.

97. If the base of a △ be given, both in magnitude and position, and the ratio of the sum of the squares of the sides to the area, the locus of the vertex is a circle.

---

* The Theorems 87 and 93–96 have been communicated to the author by Mr. W. S. M'Cay, F.T.C.D.

98. If a line of constant length slide between two fixed lines, the locus of the centre of instantaneous rotation is a circle.

99. If two sides of a $\triangle$ given in species and magnitude slide along two fixed $\odot$s, the envelope of the third side is a circle. (Bobillier).

100. If the lengths of the sides of the $\triangle$ in Ex. 99 be denoted by a, b, c, and the radii of the three $\odot$s by α, β, γ; then $a\alpha \pm b\beta \pm c\gamma =$ twice the area of the $\triangle$, the sign + or − being used according as the $\odot$s touch the sides of the $\triangle$ internally or externally.

101. If five quadrilaterals be formed from five lines by omitting each in succession, the lines of collinearity of the middle points of their diagonals are concurrent. (H. Fox Talbot.)

102. If D, D′ be the diagonals of a quadrilateral whose four sides are a, b, c, d, and two of whose opposite angles are θ, θ', then

$$D^2 D'^2 = a^2 c^2 + b^2 d^2 - 2abcd \cos(\theta + \theta').$$

103. If the sides of a $\triangle$ ABC, inscribed in a $\odot$, be cut by a transversal in the points a, b, c. If α, β, γ denote the lengths of the tangents from a, b, c to the $\odot$, then $\alpha\beta\gamma = \mathrm{A}b \, . \, \mathrm{B}c \, . \, \mathrm{C}a$.

104. If a, b, c denote the three sides of a $\triangle$, and if α, β, γ denote the bisectors of its angles,

$$\alpha\beta\gamma = \frac{8\,abc \, . \, s \, . \, \text{area}}{(a+b)(b+c)(c+a)}.$$

105. If a $\triangle$ ABC circumscribed to a $\odot$ be also circumscribed to another $\triangle$ A′B′C′, and in perspective with it, the tangents from the vertices of A′B′C′ will meet its opposite sides in three collinear points.

106. If two sides of a triangle be given in position, and its area given in magnitude, two points can be found at each of which the base subtends an angle of constant magnitude.

107. If two sides of a triangle and its inscribed circle be given in position, the envelope of its circumscribed circle is a circle.—Mannheim.

108. If the circumference of a circle be divided into an uneven number of equal parts, and the points of division denoted by the indices 0, 1, 2, 3, &c., then if the point of the circle diametrically opposite to that whose index is zero be joined with all the points in one of its semicircles, the rectangle contained by the chords terminating in the points 1, 2, 4, 8 . . . is equal to the power of the radius denoted by the number of chords.

109. A right line, which bisects the perimeter of the maximum figure contained by that perimeter, bisects also the area of the figure. Hence show, that of all figures having the same perimeter a circle has the greatest area.

110. The polar circle of a triangle, its circumscribed circle, and nine-points circle, are coaxal.

111. The polar circles of the five triangles external to a pentagon, which are fc.med by producing its sides, have a common orthogonal circle.

112. The six anharmonic ratios of four collinear points can be expressed in terms of the trigonometrical functions of an angle, namely,

$$-\sin^2\phi, \quad -\cos^2\phi, \quad \tan^2\phi, \quad -\operatorname{cosec}^2\phi, \quad -\sec^2\phi, \quad \cot^2\phi.$$

Show how to construct ϕ.

113. If the sides of a polygon of an even number of sides be cut by any transversal, the product of one set of alternate segments is equal to the product of the other set. If the number of sides of the polygon be odd, the rectangles will be equal, but will have contrary signs (CARNOT).

114. If from the angular points of a polygon of an odd number of sides concurrent lines be drawn, dividing the opposite sides each into two segments, the product of one set of alternate segments is equal to the product of the other set (PONCELET).

115. If the points at infinity on two lines divided homographically be corresponding points, the lines are divided proportionally.

116. *To construct a quadrilateral, being given the four sides and the area.*

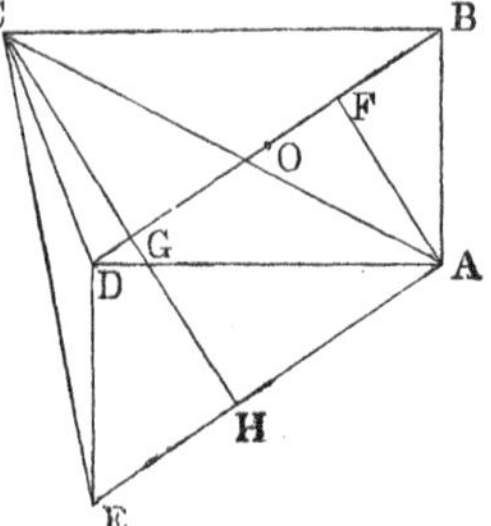

Analysis.—Let ABCD be the required quadrilateral. The four sides, AB, BC, CD, DA are given in magnitude; and the area is also given. Draw AE parallel and equal to BD. Join ED, EC; draw AF, CG perpendicular to BD; produce CG to H; bisect BD in O.

Now we have

$$BC^2 - CD^2 = 2BD \,.\, OG$$

and

$$AD^2 - AB^2 = 2BD \,.\, OF;$$

therefore

$$BC^2 + AD^2 - AB^2 - CD^2 = 2BD \,.\, FG = 2AE \,.\, AH.$$

Hence, since the four lines AB, BC, CD, DA are given in magnitude, the rectangle AE . AH is given. Now, if we suppose the line AD to be given in position, since DE is equal to AB, which is given in magnitude, the locus of the point E is a circle, and since the rectangle AH . AE is given, the locus of the point H is a circle, namely, the inverse of the locus of E.

Again, since the lines AE, AC are equal, respectively, to the diagonals of the quadrilateral, and include an angle equal to that between the diagonals, the area of the triangle ACE is equal to the area of the quadrilateral. Hence the area of the triangle ACE is given. Therefore the rectangle AE . CH is given. And it has been proved that the rectangle AE . AH is given; therefore the ratio AH : CH is given. Hence the triangle ACH is given in species. And since the point A is fixed, and H moves on a given circle, C moves on a given circle. And since D is fixed, and DC given in magnitude, the locus of the point C is another circle. Hence C is a given point.

117. Prove from the foregoing analysis that the area is a maximum when the four points A, B, C, D are concyclic.

118. In the same case prove that the angle between the diagonals is a maximum when the points are concyclic.

119. The difference of the squares of the two interior diagonals of a cyclic quadrilateral is to twice their rectangle as the distance between their middle points is to the third diagonal.

120. Inscribe in a given circle a quadrilateral whose three diagonals are given. [Make use of Ex. 119.]

121. Given the two diagonals and all the angles of a quadrilateral; construct it.

122. *If* L *be one of the limiting points of two circles,* O, O′, *and* LA, LB *two radii vectors at right angles to each other, and terminating in these circles, the locus of the intersection of tangents at* A *and* B *is a circle coaxal with* O, O′.

Dem.—Join AB, intersecting the circles again in G and H, and let fall the perpendiculars OC, O′D, LE.

Then

$AL^2 : AB . AH :: OL : OO'$ [VI. Sect. V. Prop. I. (3)].

But $AL^2 = AB.AE.$

Hence $AE : AH :: OL : OO'$;

therefore $AE : HE :: OL : O'L.$

In like manner $GE : BE :: OL : O'L.$

Hence $AG : BH :: OL : O'L,$

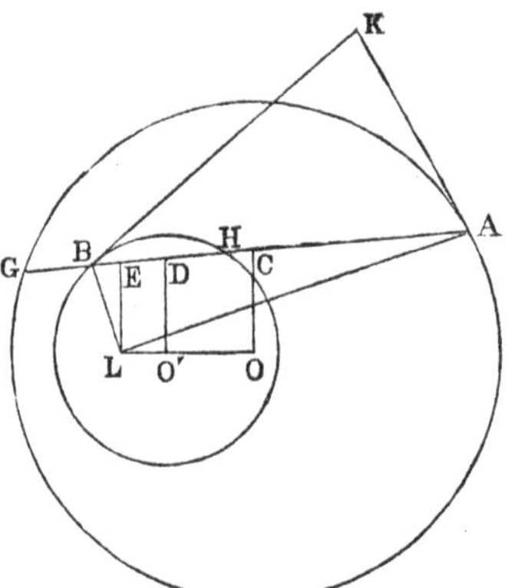

that is, in a given ratio. Therefore the tangents AK, BK are in a given ratio [EUCLID, VI. IV. Ex. 2]; and the locus of K is a circle coaxal with O, O'.

This theorem is the reciprocal of a remarkable one in Confocal Conics (see *Conics*, page 184). The demonstration of it here given, as well as that of the Proposition Ex. 116, have been communicated to me by W. S. M'CAY, F.T.C.D.

123. In the same case the locus of the point E is a circle coaxal with O, O'.

124. If O'' be the centre of the locus of E, then LO'' is half the harmonic mean between LO and LO'.

125. If r be the radius of the inscribed circle of a triangle ABC, and ρ the radius of a circle touching the circumscribed circle internally and the sides AB, AC; then $\rho \cos^2 \frac{1}{2}A = r$.

126. Prove the corresponding relation $\rho' \cos^2 \frac{1}{2}A = r'$ for the case of external contact.

127. Prove by inversion the equality of the two circles in Prop. 8, Cor. 4, p. 32.

128. If AB, CD be the diameters of two circles, and be also segments of the same line, prove that the two circles are equal which touch respectively the circles on AB, CD; their radical axis on opposite sides, and any circle whose centre is the middle point of AD.—(STEINER.)

129. Given three points, A, B, C, and three multiples, l, m, n, find a point O such that lAO + mBO + nCO may be a minimum.

130. If A, B, C, D be any four points connected by four circles, each passing through three of the points, then not only is the angle at A between the arcs ABC, ADC equal to the angle at C between CDA, CBA, but it is also equal to the angle at D between the arcs DAB, DCB; and to the angle at B between BCD, BAD.—(HAMILTON.)

131. If A, B, C be the escribed circles of a triangle, and if A′, B′, C′ be three other circles touching ABC as follows, viz. each of them touching two of the former exteriorly, and one interiorly; then A′, B′, C′ intersect in a common point P, and the lines of connexion at P with the centres of the circles are perpendicular to the sides of the triangle.

132. The line of collinearity of the middle points of the diagonals of a complete quadrilateral is perpendicular to the line of collinearity of the orthocentres of the four triangles.

133. The sines of the angles which the line of collinearity of the middle points of the diagonals of a complete quadrilateral makes with the sides are proportional to the diameters of the circles described about its four triangles.

134. If r, ρ be the radii of two concentric circles, and R the radius of a third circle (not necessarily concentric), so related to them that a triangle described about the circle r may be inscribed in R, and a quadrilateral about ρ may be inscribed in R: then

$$r/\rho + \rho/R = \rho/r.$$

135. If R, r be the radii of two circles, C, C′, of which the former is supposed to include the latter; then if a series of circles O_1, O_2, O_3, . . . O_m be described touching both and touching each other in succession, prove that if traversing the space between C, C′ n times consecutively the circle O_m touch O_1 if δ be the distance between the centres,

$$(R - r)^2 - 4Rr \tan^2 \frac{n\pi}{m} = \delta^2. \quad \text{—(Steiner.)}$$

136. If A, B, C, D be any four points, and if the three pairs of lines which join them intersect in the points b, c, d, then the nine-points circles of the four triangles ABC, ABD, ACD, BCD, and the circle about the triangle bcd, all pass through a common point P.

137. If A_1, B_1, C_1, D_1 be the orthocentres of the triangles A, B, C, &c., and b_1, c_1, d_1 the points determined by joining A_1, B_1, C_1, D_1, in pairs; then the nine-points circles of the four triangles A_1, B_1, C_1, &c., and the circumscribed circle of the triangle $b_1c_1d_1$, all pass through the former point P.—(Ex. 29.)

138. The Simson's lines (Book III. Prop. 12) of the extremities of any diameter of the circumcircle of a triangle intersect at right angles on the nine-points circle of the triangle.

139.* Every tangent to a circle is cut harmonically by the sides of a circumscribed square, and also by the sides of a circumscribed trapezoid whose non-parallel sides are equal.

140. A variable chord of a circle passes through a fixed point; its extremities and the fixed point are joined to the centre; prove that the circumcircles of the three triangles so formed touch in every position a pair of circles belonging to two given coaxal systems.

141. *Weill's Theorem.*—If two circles be so related that a polygon of n sides can be inscribed in one and circumscribed to the other, the mean centre of the points of contact is a fixed point.

142. In the same case the locus of the mean centre of any number $(n - r)$ of the points is a circle.

Weill's Theorem was published in *Liouville's Journal*, Third Series, Tome IV., page 270, for the year 1878. A proposition, of which Weill's is an immediate inference, was published by the Author in 1862, in the *Quarterly Journal of Pure and Applied Mathematics*, Vol. V., page 44, *Cor.* 2.

* Theorems 139, 140, have been communicated to the author by Robert Graham, Esq., A.M., T.C.D.

SUPPLEMENTARY CHAPTER.

RECENT ELEMENTARY GEOMETRY.

SECTION I.

Theory of Isogonal and Isotomic Points, and of Antiparallel and Symmedian Lines.

Def.—*Two lines* AX, AY *are said to be* Isogonal Conjugates, *with respect to an angle* BAC, *when they make equal angles with its bisector.*

Prop. 1.—*If from two points* X, Y *on two lines* AX, AY, which *are isogonal conjugates with respect to a given angle* BAC, *perpendiculars* XM, YN, XP, YQ *be drawn to its sides;* then—

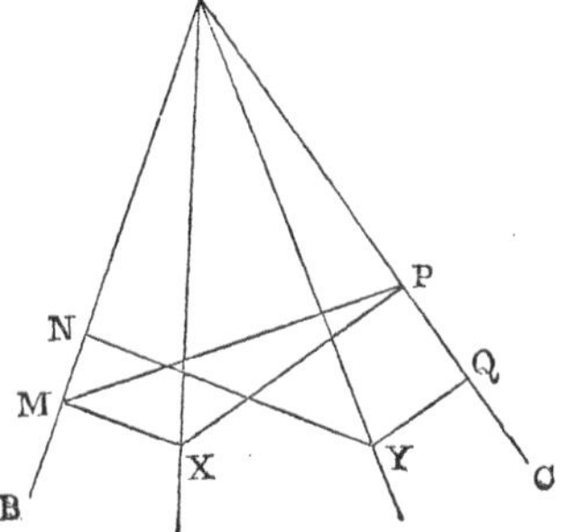

1°. *The rectangle* XM . YN = XP . YQ.

2°. *The points* M, N, P, Q *are concyclic.*

3°. MP *is perpendicular to* AY, *and* NQ *to* AX.

Dem.—1°. From the construction, we have evidently XM : AX : : YQ : AY, and AX : XP : : AY

: YN. Hence [*Euc.* V. XXII.] XM : XP : : YQ : YN; therefore XM . YN = XP . YQ.

2°. In the same manner AM : AX : : AQ : AY, and AX : AP : : AY : AN. Hence AM : AP : : AQ : AN; therefore AM . AN = AP . AQ. Hence the points M, N, P, Q are concyclic.

3°. Since the angles AMX, APX are right, AMXP is a cyclic quadrilateral; therefore the angle MAX = MPX; but MAX = YAQ. Hence MPX = YAQ, and PX is perpendicular to AQ. Hence PM is perpendicular to AY. Similarly, QN is perpendicular to AX.

Cor. 1.—*If the rectangle contained by the perpendiculars from two given points on one of the sides of a given angle be equal to the rectangle contained by the perpendiculars from the same points on the other side, the lines joining the vertex of the angle to the points are isogonal conjugates with respect to the angle.*

Prop. 2.—*The isogonal conjugates of three concurrent lines* AX, BX, CX, *with respect to the three angles of a triangle* ABC, *are concurrent.*

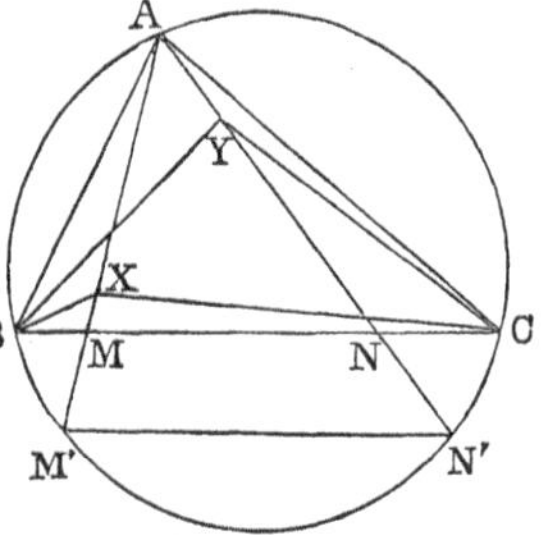

Dem.—Let the isogonal conjugates of AX, BX be AY, BY, respectively. Join CY. It is required to prove that CY is the isogonal conjugate of CX.

From 1°, Prop. 1, the rectangles of the perpendiculars from X, Y on the lines AC, BC are each equal to the rectangle contained by the perpendiculars from X, Y on AB. Hence they are equal to one another, and therefore, by Prop. 1, *Cor.* 1, the lines CX, CY are isogonal conjugates with respect to the angle C.

Def.—*The points* X, Y *are called isogonal conjugates with respect to the triangle* ABC.

Cor. 1.—*If* X, Y *be isogonal conjugates with respect to a triangle, the three rectangles contained by the distances of* X, Y *from the sides of the triangle are equal to one another.*

Cor. 2.—*The middle point of the line* XY *is equally distant from the projections of the points* X, Y *on the three sides of the triangle.*

Exercises.

1. The sum of the angles BXC, BYC is $180° + A$.

2. The line joining any two points, and the line joining their isogonal conjugates, with respect to a triangle, subtend at any vertex of the triangle angles which are either equal or supplemental.

3. $AM^2 : AN^2 :: BM . MC : BN . NC$. (Steiner.)

4. If the lines AX, AY meet the circumcircle of the triangle ABC in M′, N′, then the rectangles AB . AC, AM . AN′, and AM′. AN are equal to one another.

5. The isogonal conjugate of the point M′ is the point at infinity on the line AN′.

6. If three lines through the vertices of a triangle meet the opposite sides in collinear points, their isogonal conjugates will also meet them in collinear points.

7. If upon the sides of a triangle ABC three equilateral triangles ABC′, BCA′, CAB′ be described either externally or internally, the isogonal conjugate of the centre of perspective of the triangles ABC, A′B′C′, is a point common to the three Apollonian circles of ABC. (See *Cor.* 3, p. 86.)

8. If the lines MX, QY in fig. Prop. 1, intersect in D, and the lines MP, NQ in E, the lines AD, AE are isogonal conjugates with respect to the angle BAC.

9. If D, E be the points where two isogonal conjugates, with respect to the angle BAC, meet the base BC of the triangle BAC and if perpendiculars to AB, AC at the points B, C meet the perpendiculars to BC at D, E in the points D′, E′; D″, E″, respectively; then $BD′ . BE′ : CD″ . CE″ :: AB^4 : AC^4$.

10. In the same case $BD . BE : CD . CE :: AB^2 : AC^2$.

Def.—*The right lines* AB′, AC′ *are said to be isotomic conjugates, with respect to a side* BC, *of the tri-*

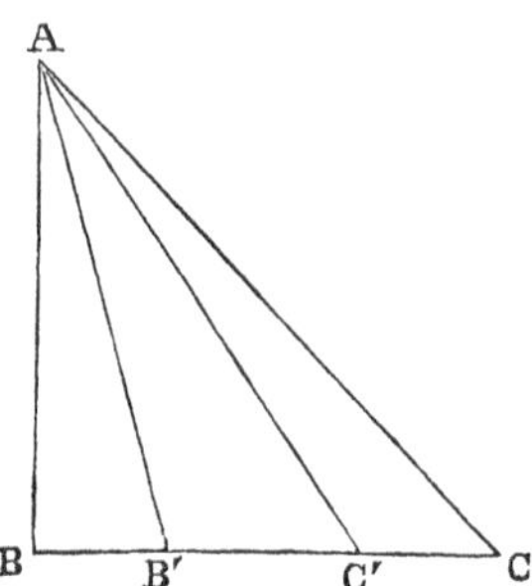

angle ABC, *when the intercepts* BB′, CC′ *on that side are equal.*

Prop. 3.—*If two points* X, Y, *in the plane of a triangle, be such that the lines* AX, AY *are isotomic with respect to the side* BC; BX, BY *with respect to* AC; *then* CX, CY *are isotomic conjugates with respect to* AB.

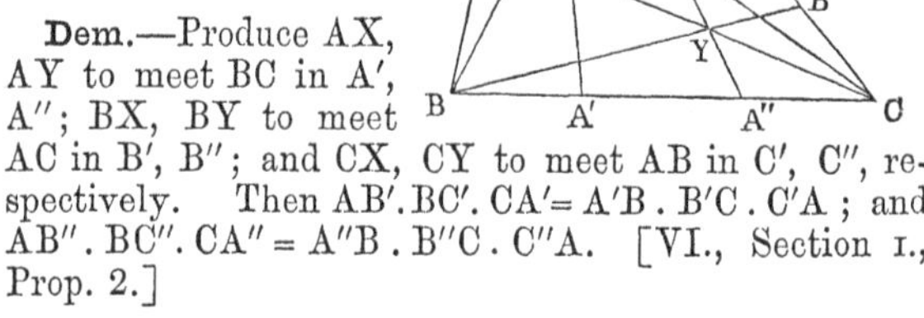

Dem.—Produce AX, AY to meet BC in A′, A″; BX, BY to meet AC in B′, B″; and CX, CY to meet AB in C′, C″, respectively. Then AB′.BC′. CA′= A′B . B′C . C′A ; and AB″. BC″. CA″ = A″B . B″C . C″A. [VI., Section I., Prop. 2.]

Hence, multiplying these equations, and omitting terms that cancel each other, we get BC′. AC″ = C″A . C′A. Hence BC″ = C′A. Q.E.D.

Def.—*Two points,* X, Y, *are said to be isotomic conjugates, with respect to a triangle* ABC, *when the pairs of lines* AX, AY; BX, BY; CX, CY *are isotomic conjugates, with respect to the sides* BC, CA, AB *respectively.*

Exercises.

1. If the multiples for which the point X is the mean centre of the points A, B, C be α, β, γ; prove that $\frac{1}{\alpha}$, $\frac{1}{\beta}$, $\frac{1}{\gamma}$ are the multiples for which the isotomic conjugate of X is the mean centre of the points A, B, C.

2. If a right line meet the sides of a triangle in the points A′, B′, C′; prove that the triads of points in which the isogonal and the isotomic conjugates of AA′, BB′, CC′, with respect to the angles A, B, C, meet the sides of the triangle, are each collinear.

Def.—*Lines* BC, B′C′ *are said to be antiparallel, with*

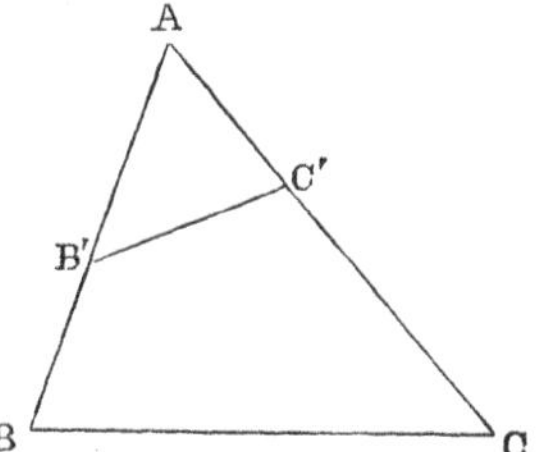

respect to the angle A, *when the angle* ABC *is equal to* AC′B′.

There are three systems of antiparallels with respect to a triangle ABC.

1°.—Antiparallels to BC with respect to the opposite angle				A.
2°.	,,	CA	,,	B.
3°.	,,	AB	,,	C.

Prop. 4.—*The antiparallels to the sides of a triangle are parallel to the tangents to its circumcircle at the angular points.*

For if LMN be the triangle formed by the tangents, the angle MAC is [*Euc.* III. XXXII.] equal to ABC; therefore MAC = AC′B′, and hence AM is parallel to C′B′.

Cor. 1—*The points* B′, C′; B, C *are concyclic.*

Cor. 2.—*The lines joining the feet of the perpendiculars of a triangle are antiparallel to its sides.*

Def.—*The isogonal conjugate of a median* AM *of a triangle is called a symmedian.*

It follows from Prop. 2 that the three symmedians of a triangle are concurrent. The point of concurrence (K) is called *the symmedian point of the triangle.*

Prop. 5.—*The perpendiculars from* K *on the sides of the triangle are proportional to the sides.*

Dem.—Let the perpendiculars from M on the sides AB, AC be MD, ME, and from K on the three sides

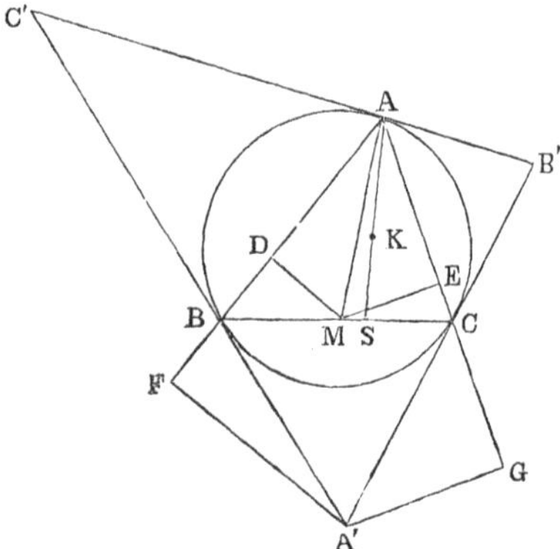

x, y, z. Then [Prop. 1, 1°] MD . z = ME . y; but MD . AB = ME . AC. Hence $y : z :: $ AC : AB, which proves the proposition.

COR. 1.— $$\frac{x}{a} = \frac{y}{b} = \frac{z}{c} = \frac{2\Delta}{a^2 + b^2 + c^2}.$$

COR. 2.—*The symmedians of a triangle are the medians of its antiparallels.*

COR. 3.—*The symmedian* AS *divides* BC *in the ratio* $AB^2 : AC^2$.

COR. 4.—*The multiples for which* K *is the mean centre of the points* A, B, C *are* a^2, b^2, c^2.

COR. 5.—*The point* K *is the isogonal conjugate of the centroid of the triangle.*

Prop. 6.—*The symmedians pass through the poles of the sides of the triangle* ABC *with respect to its circumcircle.*

Dem.—Let A′ be the pole of BC. Let fall perpendiculars A′F, A′G; then A′F : A′G : : sin A′BF : sin A′CG, or : : sin ACB : sin ABC; ∴ A′F : A′G : : AB : AC : : perpendicular from K on AB : perpendicular from K on AC. Hence the points A, K, A′ are collinear.

COR. 1.—*The polar of* K *is the axis of perspective of the triangle* ABC, *and its reciprocal, with respect to the circumcircle.*

COR 2.—*The tangent at* A *to the circumcircle, and the symmedian* AS, *are harmonic conjugates with respect to the sides* BA, AC *of the triangle.*

Prop. 7.—*The sum of the squares of the distances of* K *from the sides of* ABC *is a minimum.*

Dem.—Let x, y, z be the distances of any point whatever from the sides of ABC, and let Δ denote its area. We have the identity

$$(x^2 + y^2 + z^2)(a^2 + b^2 + c^2) - (ax + by + cz)^2$$
$$= (ay - bx)^2 + (bz - cy)^2 + (cx - az)^2;$$

but $ax + by + cz = 2\Delta$. Consequently $x^2 + y^2 + z^2$ has its minimum value when the squares which occur on the right-hand side of the identity vanish; that is, when

$$\frac{x}{a} = \frac{y}{b} = \frac{z}{c}.$$

Cor.—K *is the mean centre of the feet of its own perpendiculars on the sides of the triangle* ABC.

Def.—*If we put* $\frac{x}{a} = \frac{1}{2}\tan\omega$, ω *is called the Brocard angle of the triangle.*

Prop. 8.—$\cot\omega = \cot A + \cot B + \cot C$.

Dem.—From 5, *Cor.* 1, we have

$$\tan\omega = \frac{4\Delta}{a^2 + b^2 + c^2}.$$

Hence

$$\cot\omega = \frac{a^2 + b^2 + c^2}{4\Delta} = \frac{2bc\cos A + 2ca\cos B + 2ab\cos C}{4\Delta};$$

but

$$\frac{2bc\cos A}{4\Delta} = \cot A, \text{ \&c.}$$

Hence $$\cot\omega = \cot A + \cot B + \cot C.$$

Cor.—*If the base and the Brocard angle of a triangle be given, the locus of the symmedian point is a right line parallel to the base.*

Exercises.

1. If K_a, K_b, K_c be the points of intersection of the symmedians with the sides of the triangle ABC, the area of the triangle $K_aK_bK_c = \frac{12\Delta^3}{(a^2 + b^2 + c^2)^2}$.

2. If the side BA of the triangle ABC be produced through A until AB′ = BA; and at B′ and C be erected perpendiculars to BA, AC, respectively, meeting in I; AI is perpendicular to the symmedian passing through A.

3. If A″ be the pole of the line BC, with respect to the circumcircle A_a'', A_b'', A_c'', the feet of the perpendiculars from A″, on the sides of ABC; the area of the triangle A_a'', A_b'', A_c'' $= \dfrac{12\Delta^3}{(b^2 + c^2 - a^2)^2}$.

4. In the same case prove that the figure $A''A_b''\,A_a''\,A_c''$ is a parallelogram.

SECTION II.

Two Figures directly Similar.

Defin.—*Being given a system of points* A, B, C, D, . . . *If upon the line joining them to a fixed point O points* A′, B′, C′, &c., . . . *be determined by the conditions* $\dfrac{OA'}{OA} = \dfrac{OB'}{OB} = \dfrac{OC'}{OC} = \&c., \ldots = k$, *the two systems of points* A, B, C, &c., *and* A′, B′, C′, &c., *are said to be homothetic, and* O *is called their homothetic centre.*

Prop. 1.—1°. *The figure homothetic to a right line is a parallel right line.*

2°. *The figure homothetic to a circle is a circle.*

Dem.—1°. This follows at once from the definition and *Euc.*, VI. ii.

2°. It is evident from Book VI., Section ii., Prop. 1, *Cor.* 4.

Prop. 2.—*In two homothetic figures*—1°. *Two homologous lines are in the constant ratio k.* 2°. *Two corresponding triangles are similar.*

These are evident.

Prop. 3.—*Being given two homothetic systems, viz.,* ABC, . . . A′B′C′ . . . *If one of them,* A′B′C′, . . . *be turned round the homothetic centre O, through a constant*

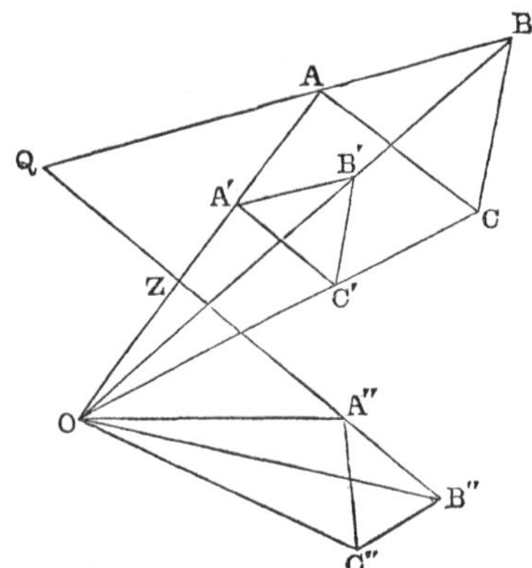

angle α, *into a new position* A″B″C″; . . . *then*—1°. *Any two homologous lines* (AB, A″B″) *are inclined at an angle* α *to each other.* 2°. *The triangles* OAA″, OBB″, &c., *are similar to each other.*

Dem. — By hypothesis, the angle OAB = OA′B′ = OA″B″, and the angle OZA″ = OZQ [*Euc.* I. xv.]. Hence [*Euc.* I. xxxii.] the angle A″OZ = ZQA. Hence ZQA = α'.

Again, the triangles OAB, OA′B′ are equiangular. Therefore OAB and OA″B″ are equiangular. Hence OA : OB : : OA″ : OB″; ∴ OA : OA″ : : OB : OB″, and the angle AOA″ = BOB″. Therefore [*Euc.* VI. vi.] the triangles AOA″, BOB″ are similar.

Cor. 1.—*Reciprocally.—If upon the lines drawn from a fixed point* O *to all the angles of a polygon* ABCD, &c., *similar triangles* OAA″, OBB″, OCC″ *be described, the polygon formed by the vertices* A″, B″, C″, &c., *is similar to the original polygon* ABCD.

Cor. 2.—*If* O *be considered as a point belonging to the*

first figure, it will be its own homologue in the second figure.

DEF.—*The point O is called a double point of the two figures; it is also called their centre of similitude.*

Prop. 4.—*Being given two polygons directly similar, it is required to find their double point.*

Let AB, A′B′ be two homologous sides of the figures; C their point of intersection. Through the two triads of points A, A′, C; B, B′, C describe two circles inter-

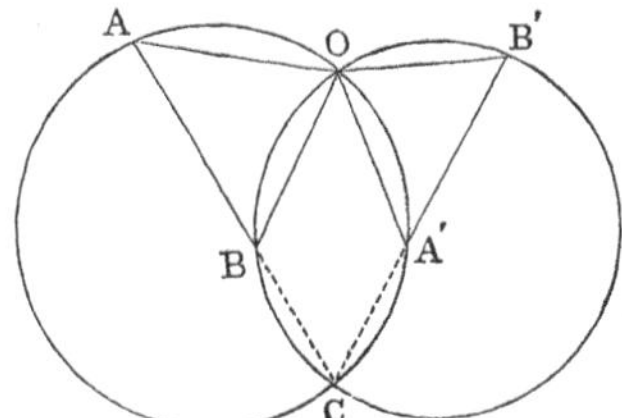

secting again in the point O: O will be the point required. For it is evident that the triangles OAB, OA′B′ are similar, and that either may be turned round the point O, so that the two bases AB, A′B′ will be parallel.

Observation.—The foregoing construction must be modified when the homologous sides of the two figures are consecutive

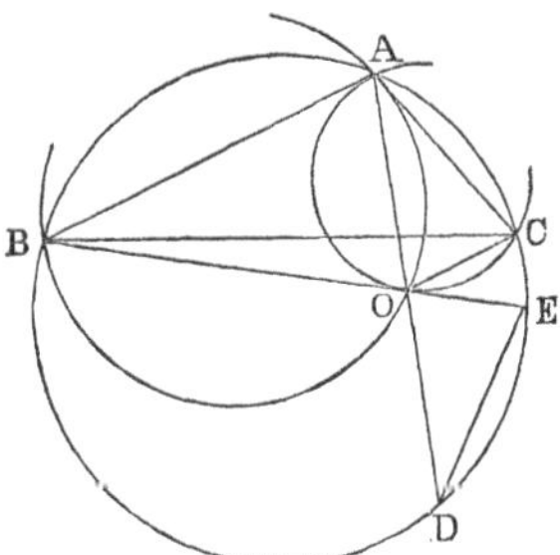

sides BA, AC of a triangle. In this case, upon the lines BA,

AC describe two segments BOA, AOC, touching AC, BC, respectively, in the point A. Then O, their point of intersection, will be the double point required, for it is evident that the triangles, BOA, AOC are directly similar.

COR. 1.—*If* AO *be produced to meet the circumcircle of the triangle* ABC *in* D, AO *is bisected in* D.

Dem.—Produce BO to meet the circle in E. Join ED. Now since the triangles BOA, AOC are directly similar, the angle OAC = ABO, and therefore = ODE. In like manner, the angle ACO = DEO. Now, because the angles DAC and ADE are equal, the arc CD = AE, Hence DE = AC; ∴ chord DE = AC; ∴ [*Euc.* I. XXVI.] DO = OA.

COR. 2.—*The distances of the double point from any two homologous points* A, A′ *are in a given ratio, because the distances are homologous lines.*

COR. 3.—*The perpendiculars drawn from the double point to any two homologous lines are in a given ratio.*

COR. 4.—*The angle subtended at the double point by the line joining two homologous points is constant.*

COR. 5.—*The line* AO *passes through the symmedian point of the triangle* BAC*; because the perpendiculars from the symmedian point on the lines* BA, AC *are proportional to these lines, and therefore proportional to the perpendiculars from* O *on the same lines.*

COR. 6.—*If* BD *be joined, the rectangle* AB . BD : AO^2 : : BC : CO.

Prop. 5.—*The centre of similitude of a given triangle* ABC, *and an equiangular inscribed triangle, is one or other of two fixed points.*

Dem.—Let DFE be an inscribed triangle, having the angles D, F, E equal to B, A, C, respectively. Then the point common to the circles BDF, AFE,

CED will [III., Section I., Prop. 17] be a given point: if this be Ω, Ω will be the centre of similitude of the triangle ABC, and an equiangular inscribed triangle, such as DFE, whose vertex F, corresponding to A, is on the line AB. In like manner, there is another point, Ω′, which is the centre of similitude of ABC, and similar inscribed triangles such as E′F′D′, having the vertex corresponding to A on the line AC.

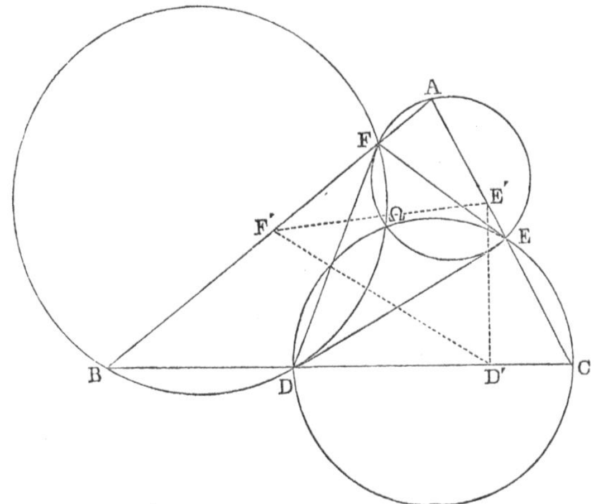

Def.—Ω, ′Ω *are called the Brocard points of the triangle* ABC.

Cor. 1.—*The circumcircle of the triangle* AΩB *touches* BC *in* B.

Dem.—Since Ω is the double point of the triangles ABC, DEF, the triangles ΩBD, ΩFA are equiangular; ∴ the angle ΩBD = ΩAF. Therefore the circle AΩB touches BC in B. In like manner, the circumcircles of the triangles BΩC, CΩA touch respectively CA in C, and AB in A.

Cor. 2.—*The three angles* ΩAB, ΩBC, ΩCA *are equal to one another, and each is equal to the Brocard angle of the triangle.*

Dem.—The angles are equal [*Euc.* III. xxxii.]. Let their common value be ω. Then, since the lines AΩ, BΩ, CΩ are concurrent, we have, from Trigonometry,

$$\sin^3\omega = \sin(A - \omega)\sin(B - \omega)\sin(C - \omega).$$

Hence $$\cot\omega = \cot A + \cot B + \cot C.$$

Therefore ω is the Brocard angle of the triangle (Section i., Prop. 8).

Exercises.

1. Inscribe in a given triangle ABC a rectangle similar to a given rectangle, and having one side on the side BC of the triangle. A is the homothetic centre of the sought rectangle, and a similar rectangle constructed on the side BC.

2. Inscribe in a given triangle a triangle whose sides will be parallel to the three given lines.

3. From the fact that a triangle ABC, and the triangle A′B′C′, whose vertices are the middle points of the sides of ABC, are homothetic; prove—1°, that the medians of ABC are concurrent; 2°, that the orthocentre, the circumcentre, and the centroid of ABC are collinear.

4. Show that Proposition 9 of Book VI., Section i., and its *Cor.* are applications of the theory of figures directly similar.

5. If figures directly similar be described on the perpendiculars of a triangle, prove that their double points are the feet of perpendiculars let fall from the orthocentre on the medians.

6. The Brocard points are isogonal conjugates with respect to the triangle BAC.

7. The system of multiples for which Ω is the mean centre of A, B, C is $\frac{1}{b^2}$, $\frac{1}{c^2}$, $\frac{1}{a^2}$; and the system for Ω' is $\frac{1}{c^2}$, $\frac{1}{a^2}$, $\frac{1}{b^2}$.

8. If the line A′B′ (Fig., Prop. 4) turn round any given point in the plane, while AB remains fixed, the locus of the double point O is a circle.

9. If through A [second Fig., Prop. 4] a line AF be drawn parallel to BC, and meeting the circle AOC again in F; prove that BF intersects the circle AOC in a Brocard point.

10. In the same Fig., if BC cut the circle AOC in G, prove that the triangles ABC, ABG have a common Brocard point.

SECTION III.

Lemoine's, Tucker's, and Taylor's Circles.

Prop. 1.—*The three parallels to the sides of a triangle through its symmedian point meet the sides in six concyclic points.*

Dem.—Let the parallels be DE′, EF′, FD′. Join ED′, DF′, FE′. Now AFKE′ is a parallelogram. AK

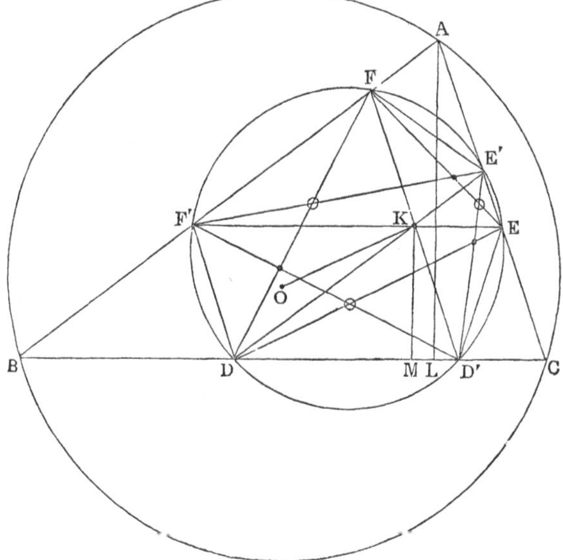

bisects FE′. Hence [Section I., Prop. 5, *Cor.* 2] FE′

is antiparallel to BC. Similarly DF′ is antiparallel to AC. Hence the angles AFE′, BF′D are equal; hence it is easy to see that FE′ is equal to F′D. In like manner it is equal to ED′.

Again, if O be the circumcentre, OA is perpendicular to FE′; therefore the perpendicular to FE′, at its middle point, passes through the middle point of KO. Hence, since FE′ = ED′ = DF′, the middle point of KO is equally distant from the six points F, E′, E, D′, D, F.

This proposition was first published in 1873, at the Congress of Lyons, *Association Francaise pour l'avancement des Sciences*, by M. Lemoine, who may be regarded as the founder of the modern Geometry of the triangle. It was rediscovered in 1883 by Mr. Tucker, *Quarterly Journal of Pure and Applied Mathematics*, p. 340.

I proved, in January, 1886 (*Proceedings* of the Royal Irish Academy), that polygons of any number of sides called harmonic polygons, can be constructed, for which a corresponding proposition is true. [See Section VI.]

Def.—*We shall call the circle through the six points* F, E′, E, D′, D, F′ *Lemoine's circle, and the hexagon of which they are the angular points Lemoine's hexagon.*

Cor. 1.—*The sides of the triangle* ABC *are divided symmetrically by Lemoine's circle.*

For it is easy to see that

$$\text{AF} : \text{FF}' :: \text{F}'\text{B} : b^2 : c^2 : a^2;$$
$$\text{BD} : \text{DD}' :: \text{D}'\text{C} : c^2 : a^2 : b^2;$$
$$\text{CE} : \text{EE}' :: \text{E}'\text{A} : a^2 : b^2 : c^2.$$

Cor. 2.—*The intercepts* DD′, EE′, FF′ *are proportional to* a^3, b^3, c^3.

Dem.—Let fall the perpendicular AL; then, since the triangles DKD′, BAC are similar,

$$\text{DD}' : x :: \text{BC} : \text{AL} :: a^2 : 2\Delta.$$

Hence $$\text{DD}' = \frac{a^2 x}{2\Delta} = \frac{a^3}{a^2 + b^2 + c^2}.$$

In like manner,

$$\text{EE}' = \frac{b^3}{a^2 + b^2 + c^2}, \quad \text{FF}' = \frac{c^3}{a^2 + b^2 + c^2}.$$

On account of this property, Mr. Tucker called the Lemoine circle "The Triplicate Ratio" circle.

Cor. 3.—*The six triangles into which the Lemoine hexagon is divided by lines from* K *to its angular points are each similar to the triangle* ABC.

Cor. 4.—*If lines drawn from the angles of a triangle* ABC, *through a Brocard point, meet the circumcircle again in* A′, B′, C′, *the figure* AB′CA′BC′ *is a Lemoine hexagon.*

Prop. 2.—*The radical axis of Lemoine's circle and the circumcircle is the Pascal's line of the Lemoine hexagon.*

Dem.—Let FE produced meet BC in X. Then since FE′ is antiparallel to BC, the points BFE′C are concyclic. Hence the rectangle BX . CX = FX . E′X. Therefore the radical axis of the Lemoine circle and the circumcircle passes through X. Hence the proposition is proved.

Cor. 1.—*The polar of the symmedian point, with respect to the Lemoine circle, is the Pascal's line of the Lemoine hexagon.*

For since DFE′D′ is a quadrilateral inscribed in the Lemoine circle, the polar of K passes through X. In like manner, it passes through each pair of intersections of opposite sides.

Cor. 2.—*If the chords* DE, D′E′ *intersect in* p, EF, E′F′ *in* q, *and* FD, F′D′ *in* r, *the triangle* pqr *is in perspective with* ABC.

Dem.—Join Aq, Cp, and let them meet in T; then denoting the perpendiculars from T on the sides of ABC by α, β, γ, respectively, we have $\alpha : \beta$: : perpendicular from p on BC : perpendicular from p on

CA—that is, : : DD′ : EE′, or : : $a^3 : b^3$. In like manner, $\beta : \gamma :: b^3 : c^3$. Hence $\alpha : \gamma :: a^3 : c^3$: : perpendicular from r on BC : perpendicular from r on AB. Hence the line Br passes through T.

Cor. 3.—*The perpendiculars from the centre of perspective of* ABC, pqr, *on the sides of* ABC *are proportional to* a^3, b^3, c^3.

Cor. 4.—*The intersections of the antiparallel chords* D′E, E′F, F′D *with Lemoine's parallels* DE′, EF′, FD′, *respectively, are collinear, the line of collinearity being the polar of* T *with respect to Lemoine's circle.*

Dem.—Let the points of intersection be P, Q, R; then CpP forms a self-conjugate triangle with respect to Lemoine's circle. Hence P is the pole of Cp. Similarly Q is the pole of Aq, and R the pole of Br; but Aq, Cp, Br are concurrent. Hence P, Q, R are collinear.

Prop. 3.—*If a triangle* $\alpha\beta\gamma$ *be homothetic with* ABC, *the homothetic centre being the symmedian point of* ABC; *and if the sides of* $\alpha\beta\gamma$ *produced, if necessary, meet those of* ABC *in the points* D, E′; E, F′; F, D′; *these six points are concyclic.*

Dem.—Let K be the symmedian point. From the hypothesis it is evident that the lines AK, BK, CK are the medians of FE′, DF′, ED′. Hence these lines are antiparallel to the sides of the triangle ABC, and therefore, as in Prop. 1, the six points are concyclic.

Cor. 1.—*The circumcentre of the hexagon* DD′EE′FF′ *bisects the distance between the circumcentres of the triangles* ABC, $\alpha\beta\gamma$.

Cor. 2.—*If the triangle* $\alpha\beta\gamma$ *vary, the locus of the circumcentre of the hexagon is the line* OK.

The circumcircles of the hexagon, when the triangle $\alpha\beta\gamma$ varies, were first studied by M. Lemoine at the Congress of Lyons, 1873. Afterwards by Neuberg see *Mathesis*, vol. i., 1881, pp. 185–190; by M'CAY, *Educational Times*, 1883, Question 7551; by TUCKER, *Quarterly Journal of Pure and Applied Mathematics*, vol. xx., 1885, pp. 57–59. Neuberg has called them TUCKER'S CIRCLES.

COR. 3.—*If the triangle* $\alpha\beta\gamma$ *reduce to the point K, the* TUCKER'S CIRCLE, *whose centre is the middle point of* OK, *is* LEMOINE'S CIRCLES.

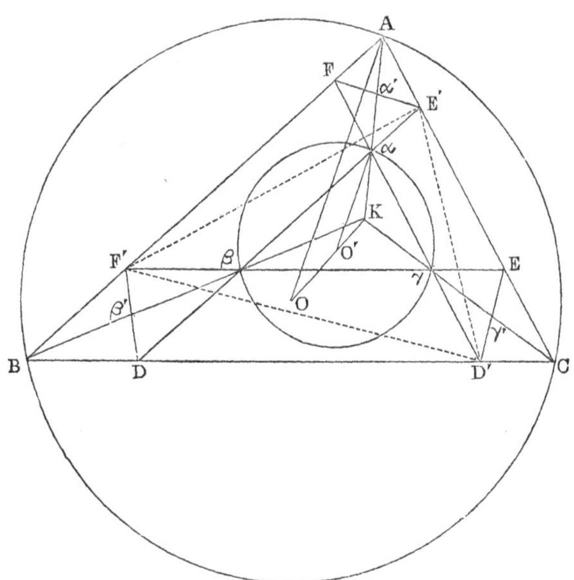

COR. 4.—*If parallels to the sides af the orthocentre triangle pass through* K, *the centre of the* TUCKER'S *circle will be* K, *the inscribed triangles will have their sides perpendicular to those of* ABC, *and the intercepts which the circle makes on the sides of* ABC *will be proportional to the cosines of its angles. This is called the* COSINE CIRCLE.

COR. 5.—*The centre of perspective of* ABC, *and the triangle formed by the points* F, D, E, *is the Brocard point* Ω, *and of* ABC *and* E′F′D′ *is* Ω'.

COR. 6.—*If* O_1 *be the centre of a Tucker's circle, and* R_1 *its radius*, $O\Omega : O_1\Omega :: R : R_1 :: O\Omega' : O_1\Omega'$.

Prop. 4.—If A′, B′, C′ *be the vertices of the orthocentric triangle of* ABC, *and* K, N; K′, N′; K″, N″ *their projections on the sides, these projections are concyclic.*

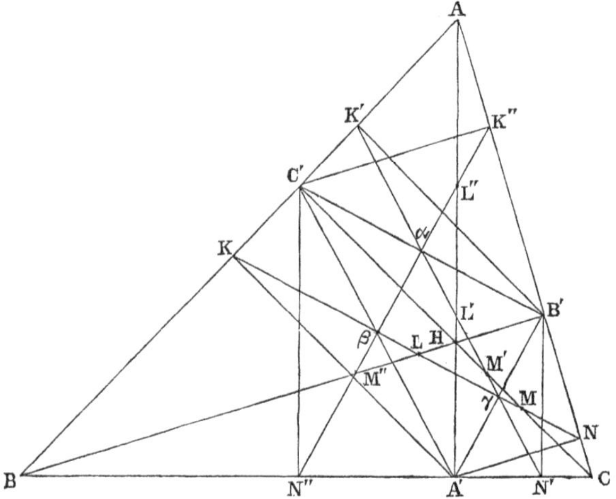

Dem.—Let H be the orthocentre. Then the figures AKA′N, AC′HB′ are homothetic, A being the homothetic centre. Hence KN is parallel to C′B′, and evidently K′K″ is antiparallel to C′B′; $\therefore$ K′K″ is antiparallel to KN. Hence the points KK′K″N are concyclic. Similarly K′K″KN′ are concyclic. Hence the proposition is proved.

This circle was first discussed in England by H. M. Taylor in a Paper published in the *Proc.* of the London Mathematical Society, vol. xv., p. 122. It is called the Taylor circle of the triangle. I shall denote it by the letter T, and the Taylor circles of the triangles BHC, CHA, AHB, respectively, by T_1, T_2, T_3.

Prop. 5.—*The chords* KN, K′N′, K″N″ *of* T *meet the sides of the triangles* BHC, CHA, AHB, *respectively, in their points of intersection with the Taylor's circles of these triangles.*

Dem.—Let KN meet BH in L, and HC in M. Now it is evident that the point A′ is common to the circum-circles of the four triangles formed by the lines AB, AC, BB′, CC′. Hence [Book III., Prop. 12, *Cor.* 2] the projections of A on these lines are collinear; therefore the points L, M are the projections of A′ on BH, HC, respectively. Similarly M′ is the projection of B′ on HC, and M″ of C′ on BH. Therefore the circle T_1 passes through the points L, M; M′, N′; M″, N″; that is, through the points of intersection of KN, K′N′, K″N″, with the sides of the triangle BHC. Hence the proposition is proved.

Prop. 6.—*The centres of* T, T_1, T_2, T_3 *coincide respectively with the incentre and the excentres of the triangle formed by joining the middle points of the sides of the orthocentric triangle of* ABC.

Dem.—The line KN is evidently the Simson's line of the point A′ with respect to the triangle BHC′, and C′ is the orthocentre of BHC′. Hence A′C′ is bisected by KN [Book III., Prop. 14]. Similarly, KN bisects A′B′, therefore it bisects two of the sides of the triangle A′B′C′, and similar properties hold for K′N′, and K″N. Hence, if α, β, γ be the middle points of the sides of A′B′C′, each of the lines KN, K′N′, K″N″ passes through two of these points. Again, since B′C′ is bisected in α, the triangle αN′N″ is isosceles, and the bisector of the angle α bisects N′N″ perpendicularly, and therefore passes through the centre of T. Similarly, the bisectors of the other angles of the triangle $\alpha\beta\gamma$ pass through the centre of T. Therefore the centre of T is the incentre of the triangle $\alpha\beta\gamma$. Similarly, the excentres of $\alpha\beta\gamma$ are the centres of T_1, T_2, T_3.

COR. 1.—*Taylor's circle* T *is one of the Tucker system of the triangle* ABC.

For, if we consider the triangle KK″N′ inscribed in ABC, the angle KK″N′ is equal to KK′N′, since the points K, K′, K″, N are concyclic; but KK′N′ is equal to C, since K′N′ is antiparallel to AC. Hence KK″N is equal to C. Again, N′KK″ is equal to N′K′K″, which, since K′K″ is parallel to BC, is equal to K′N′B, and therefore equal to A. Therefore KK″N′ is similar to ABC. Hence its circumcircle T is one of the Tucker system of the triangle ABC.

COR. 2.—*The radical axes of the circles* T, T_1, T_2, T_3 *taken in pairs are the sides and the altitudes of the triangle* ABC.

COR. 3.—*The figure formed by the centres of* T, T_1, T_2, T_3 *is similar to, and in perspective with, that formed by the points* H, A, B, C.

For, H, A, B, C are the incentre and the excentres of the triangle A′B′C′, which is similar to, and in perspective with, $\alpha\beta\gamma$.

Prop. 7.—*Taylor's circle* T *cuts orthogonally the three escribed circles of the orthocentric triangle of* ABC, *and each of the circles* T_1, T_2, T_3 *cuts orthogonally the inscribed and two of the escribed circles of the same triangle.*

Dem.—Let the perpendiculars from A, B, C on the lines B′C′, C′A′, A′B′, respectively, be π_1, π_2, π_3, respectively; then π_1, π_2, π_3 are the radii of the escribed circles of A′B′C′. Now, since the triangles AB′C′, ABC are similar, $\pi_1^2 : AA'^2 :: AC'^2 : AC^2$; that is, $\pi_1^2 : AN \cdot AC :: AC \cdot AK'' : AC^2$; $\therefore \pi_1^2 = AN \cdot AK''$; but AN . AK″ is equal to the square of the tangent from A to T. Hence the circle whose centre is A, and radius

π_1, cuts the circle T orthogonally, and similarly the circles whose centres B, C, and radii π_2, π_3 cut T orthogonally. Hence the proposition is proved.

Prop. 8.—*If σ be the semiperimeter of the triangle $\alpha\beta\gamma$. and ρ, ρ_1, ρ_2, ρ_3 the radii of its incircle and circumcircles; then the squares of the radii of Taylor's circle are, respectively,* $\rho^2 + \sigma_1^2$, $\rho_1^2 + (\sigma - \alpha)^2$, $\rho_2^2 + (\sigma - \beta)^2$, $\rho_3^2 + (\sigma - \gamma)^2$.

Dem.—Since the triangle A′N″C′ is right-angled at N″, and A′C′ is bisected in β, N″β is equal to A′β; that is [*Euc.* I. xxxiv.], equal to $\alpha\gamma$. In like manner, αK″ is equal to $\beta\gamma$. Hence N″K″ = 2σ; and since the circle T passes through the points N″, K″, and is concentric with the incircle of $\alpha\beta\gamma$, we have the square of the radius of T = $\rho^2 + \frac{1}{4}\text{N}''\text{K}''^2 = \rho^2 + \sigma^2$.

Again, if M″C′ be joined, the figure C′M″B′K″ is a rectangle. Hence M″K″ = 2αB′ = $2\beta\gamma = 2\alpha$, but N″K″ = 2σ; $\therefore$ N″M″ = $2(\sigma - \alpha)$, and, as before, the square of the radius of $\text{T}_1 = \rho_1^2 + (\sigma - \alpha)^2$. Hence the proposition is proved.

Cor.—*The sum of the squares of the radii of Taylor's circles is equal to the square of the diameter of the circumcircle.*

For it is easy to see that the squares of the radii of the four circles are, respectively, equal to,

$$4\text{R}^2(\sin^2\text{A}\ \sin^2\text{B}\ \sin^2\text{C} + \cos^2\text{A}\ \cos^2\text{B}\ \cos^2\text{C}),$$
$$4\text{R}^2(\cos^2\text{A}\ \sin^2\text{B}\ \sin^2\text{C} + \sin^2\text{A}\ \cos^2\text{B}\ \cos^2\text{C}),$$
$$4\text{R}^2(\sin^2\text{A}\ \cos^2\text{B}\ \sin^2\text{C} + \cos^2\text{A}\ \sin^2\text{B}\ \cos^2\text{C}),$$
$$4\text{R}^2(\sin^2\text{A}\ \sin^2\text{B}\ \cos^2\text{C} + \cos^2\text{A}\ \cos^2\text{B}\ \sin^2\text{C}),$$

and the sum of these is 4R^2.

Exercises.

1. The chords DE, EF, FD of Lemoine's hexagon meet the chords F′D′, D′E′, E′F′, respectively in three points forming a triangle homothetic with ABC.

2. The triangle formed by the three alternate sides DF′, FE′, ED′, produced, is homothetic with the orthocentric triangle, and their ratio of similitude is 1 : 4 cos A cos B cos C.

3. The system of circles which are circles of similitude of the circumcircle and Tucker's circles, respectively, are coaxal.

4. The perpendiculars from K on the sides of the triangle Ex. 2 are proportional to a^2, b^2, c^2.

5. The perimeter of Lemoine's hexagon is $\dfrac{a^3 + b^3 + c^3 + 3abc}{a^2 + b^2 + c^2}$, and its area $\dfrac{\Delta(a^4 + b^4 + c^4 + a^2b^2 + b^2c^2 + c^2a^2)}{(a^2 + b^2 + c^2)^2}$.

6. If the cosine circle intersect the sides of ABC in the points D, D′, E′, E′, F, F′, the figures DD′E′F, EE′F′D, FF′D′E are rectangles; and their areas are proportional to sin 2A, sin 2B, sin 2C.

7. In the same case, the diagonal of each rectangle passes through the symmedian point. This affords a proof of the theorem, that the middle point of any side, the middle point of its corresponding perpendicular, and the symmedian point, are collinear.

8. If the sides of the triangle $\alpha\beta\gamma$ (fig., Prop. 3) produced, if necessary, meet the tangents at A, B, C to the circumcircle, six of the points of intersection are concyclic, and three are collinear.

9. If the distance OK between the circumcentre and symmedian point be divided in the ratio $l : m$ by the centre of one of Tucker's circles, and if R, R′ be the radii of the circumcircle and the cosine circle, the radius of Tucker's circle is $\dfrac{\sqrt{l^2R'^2 + m^2R^2}}{l + m}$.

10. The square of the diameter of Lemoine's circle is $R^2 + R'^2$.

11. If a variable triangle $\alpha\beta\gamma$ of given species be inscribed in a fixed triangle ABC, and if the vertices of $\alpha\beta\gamma$ move along the sides of ABC, the centre of similitude F of $\alpha\beta\gamma$, in any two of its positions, is a fixed point. (Townsend.)

12. In the same case, if the circumcircle of $\alpha\beta\gamma$ meet the sides of ABC in the three additional points α', β', γ'; the triangle $\alpha'\beta'\gamma'$ is given in species, and the centre of similitude F′ of it, in any two of its positions, is a fixed point. (Taylor.)

13. Also F, F′ are isogonal conjugates with respect to the triangle.

14. The locus of the centre of the circle $\alpha\beta\gamma$ is a right line.

15. If through the Brocard points and the centre of any of Tucker's circles a circle be described, cutting Tucker's circle in X, Y; prove $\Omega X + \Omega' X = \Omega Y + \Omega' Y =$ constant.

16. The locus of the inverse of either Brocard point with respect to a Tucker circle is a right line.

17. If the middle points of the lines AH, BH, CH be A″, B″, C″, respectively, and the middle points of the sides BC, CA, AB be A‴, B‴, C‴; then the Simson's line of any of these six points, with respect to the triangle A′B′C′, passes through the centres of two of Taylor's circles.

18. If the orthocentres of the triangles AB′C′, HB′C′ be P, Q, respectively, the lines A′P, A′Q are bisected by the centres of two of Taylor's circles.

19. The Simson's lines of any vertex of the triangle A′B′C′, with respect to the four triangles A″B″C‴, B″C″A‴, C″A″B‴, A‴B‴C‴ pass respectively through the centres of Taylor's circles.

20. Prove that the intercept which the loci in Ex. 16 make on any side of the triangle subtends a right angle at either Brocard point.

SECTION IV.

General Theory of a System of Three Similar Figures.

Notation.—Let F_1, F_2, F_3 be three figures directly similar; a_1, a_2, a_3 three corresponding lengths; α_1 the constant angle of intersection of two corresponding lines of F_2 and F_3; α_2, α_3 the angles of two corresponding lines of F_3 and F_1, of F_1 and F_2, respectively; S_1 the double point of F_2 and F_3; S_2 that of F_3 and F_1; S_3 that of F_1 and F_2. We shall denote also by (O, AB) the distance from the point O to the line AB.

Def. 1.—*The triangle formed by the three double points* S_1, S_2, S_3 *is called the triangle of similitude of* F_1, F_2, F_3; *and its circumcircle their circle of similitude.*

Prop. 1.—*In every system of three figures directly similar, the triangle formed by three homologous lines is in perspective with the triangle of similitude, and the locus of the centre of perspective is their circle of similitude.*

Dem.—Let d_1, d_2, d_3 be three homologous lines forming the triangle $D_1D_2D_3$; we have, by hypothesis,

$$\frac{(S_1, d_2)}{(S_1, d_3)} = \frac{a_2}{a_3}; \quad \frac{(S_2, d_3)}{(S_2, d_1)} = \frac{a_3}{a_1}; \quad \frac{(S_3, d_1)}{(S_3, d_2)} = \frac{a_1}{a_2}.$$

Hence it follows that the lines S_1D_1, S_2D_2, S_3D_3 co-intersect in a point K, whose distances from the lines d_1, d_2, d_3 are proportional to a_1, a_2, a_3. The triangle $D_1D_2D_3$ being given in species, its angles are the supplements of α_1, α_2, α_3. Hence the angles D_1KD_2, D_2KD_3, D_3KD_1,

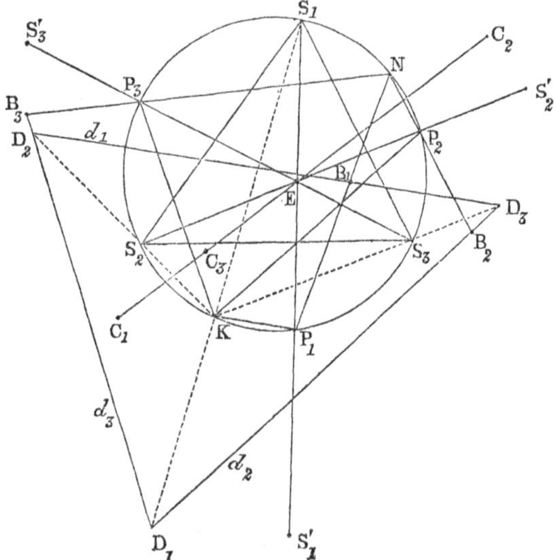

are constants; that is, the angles S_1KS_2, S_2KS_3, S_3KS_1 are constants. Hence the point K moves on three circles passing through S_1 and S_2, S_2 and S_3, S_3 and S_1; that is, it moves on the circumcircle of the triangle $S_1S_2S_3$.

Def. 2.—*The point* K *is called the perspective centre of the triangle* $D_1D_2D_3$.

Prop. 2.—*In every system of three similar figures there is an infinite number of triads of concurrent homologous lines. These lines turn round three fixed points* P_1, P_2, P_3 *of the circle of similitude, and their point of concurrence is on the same circle.*

Dem.—Let K be the perspective centre of a triangle $D_1 D_2 D_3$ formed by three homologous lines. Through K draw three parallels, KP_1, KP_2, KP_3 to the sides of $D_1D_2D_3$. These are three homologous lines.

For $$\frac{(S_1, KP_2)}{(S_1, KP_3)} = \frac{(S_1, d_2)}{(S_1, d_3)} = \frac{a_2}{a_3}, \text{ \&c.}$$

The point P_1 is fixed; for the angle S_1KP_1 is equal to the inclination KD_1 to D_2D_3, which is constant. Hence the arc SP_1 and, therefore, the point P_1 is given Similarly the points P_2, P_3 are fixed.

Def. 3.—*The points* P_1, P_2, P_3 *are called the invariable points, and the triangle* $P_1P_2P_3$ *the invariable triangle.*

Cor. 1.—*The invariable triangle is inversely similar to the triangle formed by three homologous lines.*

For the angle $P_2P_3P_1 = P_2KP_1 =$ angle $D_1D_3D_2$, and similarly for the other angles.

Cor. 2.—*The invariable points form a system of three corresponding points.*

For the angle
$$P_2S_1P_3 = a_1, \text{ and } \frac{S_1P_2}{S_1P_3} = \frac{(S_1, KP_2)}{(S_1, KP_3)} = \frac{a_2}{a_3}.$$

Cor. 3.—*The lines of connexion of the invariable points* P_1, P_2, P_3, *to any point whatever* (K) *of the circle of similitude, are three corresponding lines of the figures* F_1, F_2, F_3.

In fact these lines pass through three homologous points, P_1, P_2, P_3, and make with each other, two by two, angles equal to α_1, α_2, α_3.

Prop. 3.—*The triangle formed by any three corresponding points is in perspective with the invariable triangle, and the locus of their centre of perspective is the circle of similitude.*

Dem.—Let B_1, B_2, B_3 be three corresponding points; then P_1B_1, P_2B_2, P_3B_3 are three corresponding lines; and since they pass through the invariable points they are concurrent, and their point of concurrence is on the circle of similitude. Hence the proposition is proved.

Prop. 4.—*The invariable triangle and the triangle of similitude are in perspective, and the distances of their centre of perspective from the sides of the invariable triangle are inversely proportional to* a_1, a_2, a_3.

Dem.—We have

$$\frac{a_2}{a_3} = \frac{S_1P_2}{S_1P_3} = \frac{(S_1, P_1P_2)}{(S_1, P_1P_3)}; \quad \frac{a_3}{a_1} = \frac{(S_2, P_2P_3)}{(S_2, P_2P_1)}; \quad \frac{a_1}{a_2} = \frac{(S_3, P_3P_1)}{(S_3, P_3P_2)}.$$

Hence the lines S_1P_1, S_2P_2, S_3P_3 are concurrent.

DEF. 4.—*The centre of perspective of the invariable triangle, and the triangle of similitude, is called the director point of the three similar figures* F_1, F_2, F_3.

Prop. 5.—

Let S_1' *be the point of* F_1, *which is homologous to* S_1, *considered in* F_2 *and* F_3.

Let S_2' *be the point of* F_2, *which is homologous to* S_2, *considered in* F_3 *and* F_1.

Let S_3' *be the point of* F_3, *which is homologous to* S_3, *considered in* F_1 *and* F_2.

The triangle $S_1'S_2'S_3'$ *is in perspective both with the invariable triangle and with the triangle of similitude; and the three triangles have a common centre of perspective.*

Dem.—By hypothesis, the three points S_1', S_1, S_1 are homologous points of the figures F_1, F_2, F_3. Hence the lines $S_1'P_1$, S_1P_2, S_1P_3 are concurrent. Hence the points S_1', P_1, S_1 are collinear. Similarly S_2', P_2, S_2 are collinear, and S_3', P_3, S_3 are collinear. *Hence the proposition is proved.*

DEF. 5.—*The points* S_1', S_2', S_3' *are called the adjoint points of the figures.*

Prop. 6.—*In three figures,* F_1, F_2, F_3, *directly similar, there exists an infinite number of systems of three corresponding points which are collinear. Their loci are three circles, each passing through two double points and through* E, *the centre of perspective of the triangle of similitude, and the invariable triangle. Also the line of collinearity of each triad of corresponding points passes through* E.

Dem.—Let C_1, C_2, C_3 be three homologous collinear points. Since S_2 is the double point of the figures F_3, F_1; the triangles $S_2C_3C_1$, $S_2P_3P_1$ are similar; therefore the angle $S_2C_3C_1$ is equal to the angle $S_2P_3P_1$, and therefore [*Euc.* III., XXI.] equal to the angle S_2S_1E. In like manner, the angle $C_2C_3S_1$ is equal to S_1S_2E; therefore the angle $S_2C_3S_1$ is equal to S_2ES_1. Hence the locus of C_3 is the circumcircle of the triangle S_1ES_2.

Again, since $S_2C_3ES_1$ is a cyclic quadrilateral, the angles S_2S_1E, EC_3S_2 are supplemental. Hence the angles $S_2C_3C_1$, EC_3S_2 are supplemental; therefore the points C_1, C_3, E are collinear, and the proposition is proved.

COR. 1.—*The circumcircle of the triangle* S_1ES_2 *passes through* S_3'. For S_3' is a particular position of C_3.

COR. 2.—*The lines* C_1P_1, C_2P_2, C_3P_3 *are concurrent, and the locus of their point of concurrence is the circle of similitude.*

The substance of this Section is taken from *Mathesis*, vol. ii., page 73. Propositions 1–5 are due to M. G. TARRY, and Proposition 6 to NEUBERG.

Exercises.

1. If in the invariable triangle be inscribed triangles equiangular to the triangle of similitude, so that the vertex corresponding to S_1 will be on the side P_2P_3, &c., the centre of similitude of the inscribed triangles is the director point.

2. If V_1, V_2, V_3 be the centres of the circles which are the loci of the points C_1, C_2, C_3; then the sum of the angles P_1, S_1, V_1 is equal to the sum of P_2, S_2, V_2, equal to the sum of P_3, S_3, V_3, equal to two right angles.

3. The system of multiples for which the director point is the mean centre of the invariable points is $a_1 \operatorname{cosec} \alpha_1$, $a_2 \operatorname{cosec} \alpha_2$, $a_3 \operatorname{cosec} \alpha_3$.

4. The director point, and either the triangle of similitude or the invariable triangle suffice to determine the figures F_1, F_2, F_3.

5. Prove that the triangles $S_1S_2S_3'$, $S_2S_3S_1'$, $S_3S_1S_2'$ are similar.

6. If $S_1'S_2$, $S_2'S_1$ meet in S_3'', prove that the triangle $S_1S_2S_3''$, and the two other analogous triangles $S_2S_3S_1''$, $S_3S_1S_2''$, are similar.

SECTION V.

SPECIAL APPLICATION OF THE THEORY OF FIGURES DIRECTLY SIMILAR.

1°. The Brocard circle.

DEF. 1.—*If* O *be the circumcentre, and* K *the symmedian point of the triangle* ABC, *the circle on* OK *as diameter is called the Brocard circle of the triangle.*

DEF. 2.—*If from* O *perpendiculars be drawn to the sides of the triangle* ABC, *these meet the Brocard circle in three other points* A′, B′, C′, *forming a triangle, which we shall call Brocard's First Triangle.*

The Brocard circle is called after M. H. Brocard, *Chef de Bataillon*, who first studied its properties in the Nouvelle Correspondance, Mathematique, tomes III., IV., V., VI. (1876, '77, '78, '79); and subsequently in two Papers read before the Association Française pour l'avancement des Sciences, Congrès d'Alger, 1881, and Congrès de Rouen, 1883. Several Geometers have since studied its properties, especially Neuberg, M'Cay, and Tucker.

Prop. 1.—*Brocard's first triangle is inversely similar to* ABC.

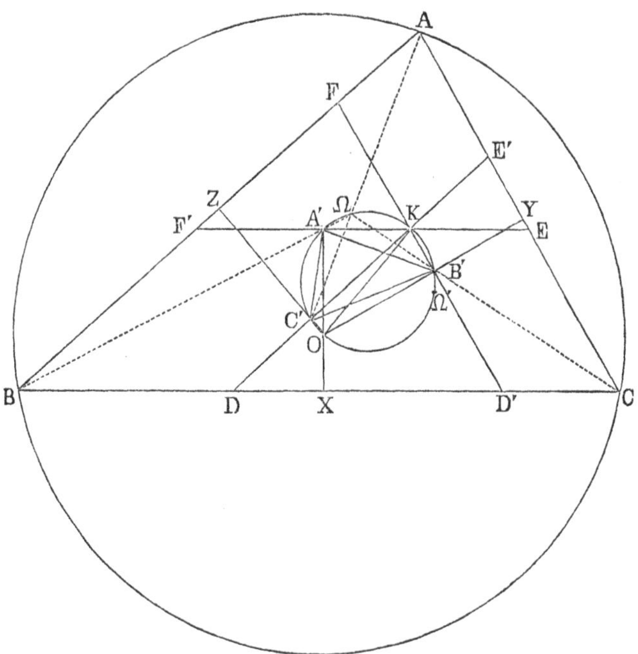

Dem.—Since OA′ is perpendicular to BC, and OB′ to AC, the angle A′OB′ is equal to ACB; but [*Euc.* III. XXI.] A′OB′ is equal to A′C′B′. Hence A′C′B′ is equal to ACB. In like manner, the other angles of these triangles are equal; and since they have different aspects, they are inversely similar.

Cor.—*The three lines* A′K, B′K, C′K, *produced, coincide with Lemoine's parallels. For since the angle* O′AK *is right,* A′K *is parallel to* BC.

Prop 2.—*The three lines* A′B, B′C, C′A *are concurrent, and meet on the Brocard circle, in one of the Brocard points.*

Dem.—Produce BA′, CB′ to meet in Ω. Then since the perpendiculars from K, on the sides of ABC, are proportional to the sides, and these perpendiculars are equal, respectively, to A′X, B′Y, C′Z, the triangles BA′X, CB′Y, A′CZ are equiangular; ∴ the angle BA′X is equal to CB′Y, or [*Euc.* I. xv.] equal to ΩB′O. Hence the points A′, Ω, B′, O are concyclic, and ∴ BA′, CB′ meet on the Brocard circle. In like manner, BA′, AC′ meet on the Brocard circle. Hence the lines A′B, B′C, C′A are concurrent, and evidently (Section II., Prop. 5) the point of concurrence is a Brocard point. In the same manner it may be proved that the three lines AB′, BC′, C′A meet on the Brocard circle in the other Brocard point.

Prop. 3.—*The lines* AA′, BB′, CC′ *are concurrent.*

Dem.—Since Lemoine's circle, which passes through F′ and E, and Brocard's circle, which passes through A′ and K, are concentric, the intercept F′A′ is equal to KE. Hence the lines AA′, AK are isotomic conjugates with respect to the angle A. In like manner, BB′, BK are isotomic conjugates with respect to the angle B, and CC′ and CK with respect to C. Therefore the three lines AA′, BB′, CC′, are concurrent: their point of concurrence is the isotomic conjugate of K with respect to the triangle ABC.

Cor. 1.—*The Brocard points are on the Brocard circle.*

Cor. 2.—*The sides of the triangle* FDE *are parallel to the lines* AΩ, BΩ, CΩ, *respectively.*

Dem.—Join DF. Then since AF = KE′; but KE′ = DC′; ∴ AF = DC′. Hence [*Euc.* I. xxxiv.] DF is parallel to AC′—that is, to AΩ, &c.

In the same manner it may be proved that the sides of D′E′F are parallel to AΩ′, BΩ′, CΩ′, respectively.

Cor. 3.—*The six sides of Lemoine's hexagon, taken in order, are proportional to* sin (A − ω), sin ω, sin (B − ω), sin ω, sin (C − ω), sin ω.

Cor. 4.—Ω *and* K *are the Brocard points of the triangle* DEF, *and* Ω′ *and* K *of* D′E′F′.

Cor. 5.—*The lines* AA′, BB′, CC′ *are isogonal conjugates of the lines* A*p*, B*q*, C*r* (Section ii., Prop. 2, Cor. 2) *with respect to the triangle* ABC.

Def.—*If the Brocard circle of the triangle* ABC *meet its symmedian lines in the points* A″, B″, C″, *respectively,* A″B″C″ *is called Brocard's second triangle.*

Prop. 4.—*Brocard's second triangle is the triangle of similitude of three figures, directly similar, described on the three sides of the triangle* ABC.

Dem.—Since OK (fig., Prop. 5) is the diameter of the Brocard circle, the angle OA″K is right. Hence A″ is the middle point of the symmedian chord AT, and is therefore [Section ii., Prop. 4, *Cors.* 1, 5] the double point of figures directly similar, described on the lines BA, AC. Hence [Section iv., Def. 1] the proposition is proved.

Prop. 5.—*If figures directly similar be described on the sides of the triangle* ABC, *the symmedian lines of the triangle formed by three corresponding lines pass through the vertices of Brocard's second triangle.*

Dem.—Let bac be a triangle formed by three corresponding lines, then bac is equiangular to BAC; and since A″ is the double point of figures described on BA, AC, and ba, ac are corresponding lines in these figures, the line A″a divides the angle bac into parts respectively equal to those into which A″A divides the angle BAC. Hence A″a is a symmedian line of bac, and similarly B″b, C″c are symmedian lines of the same triangle.

Cor. 1.—*The symmedian point of the triangle* bac *is on the Brocard circle of* BAC.

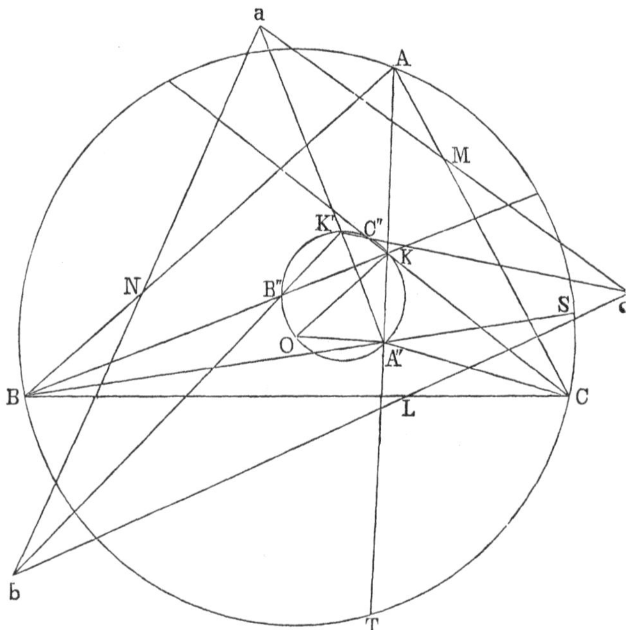

Dem.—Because the triangle bac is formed by three homologous lines, and A″B″C″ is the triangle of similitude, and [Section iv., Prop. 1] these are in perspective; therefore their centre of perspective, K′, is a point on the circle of similitude, that is, on the Brocard circle.

Cor. 2.—*The vertices* A′, B′, C′ *of Brocard's first triangle are the invariable points of the three figures directly similar, described on the sides of* BAC.

For the angle KA′K′ is equal to KA″K′, and that is evidently equal to CLc from the properties of the similitude of BAC, bac; but A′K is parallel to LC. Hence A′K′ is parallel to bc. In like manner, B′K′, C′K′ are parallel to ac, ab, respectively. Hence A′K′, B′K′, C′K′, form a system of three corresponding lines, and A′, B′, C′ are the invariable points.

Cor. 3.—*The centre of similitude of the triangles* bac, BAC *is a point on the Brocard circle.*

For since the figures K′bac, KBAC are similar, and K′a, KA are corresponding lines of these figures intersecting in A″, the centre of similitude [Section II., Prop. 4] is the point of intersection of the circum-circles of the triangles A″aA, A″KK′; but one of these is the Brocard circle. Hence, &c.

Cor. 4.—*In like manner, it may be shown that the centre of similitude of two figures, whose sides are two triads of corresponding lines of any three figures directly similar, is a point on the circle of similitude of the three figures.*

Cor. 5.—*If three corresponding lines be concurrent, the locus of their point of concurrence is the* Brocard Circle.

This theorem, due to M. Brocard, is a particular case of the theorem Section IV., Prop. 2, or of Cor. 1, due to M'Cay, or of either of my theorems, Cors. 3, 4.

2°. The Nine-points Circle.

6. Let ABC be a triangle, whose altitudes are AA′, BB′, CC′; the triangles AB′C′, A′BC′, A′B′C are inversely similar to ABC. Then if we consider these triangles as portions of three figures, directly similar,

F_1, F_2, F_3, we have three triads of homologous points,

	F_1,	F_2,	F_3.
First triad,	A,	A′,	A′;
Second ,,	B′,	B,	B′;
Third ,,	C′,	C′,	C.

The double points A′, B′, C′ are the feet of the perpendiculars. The three homologous lines, AB′, A′B, A′B′, equal to AB cos A, AB cos B, AB cos C. Hence the three homologous lines are proportional to cos A, cos B, cos C.

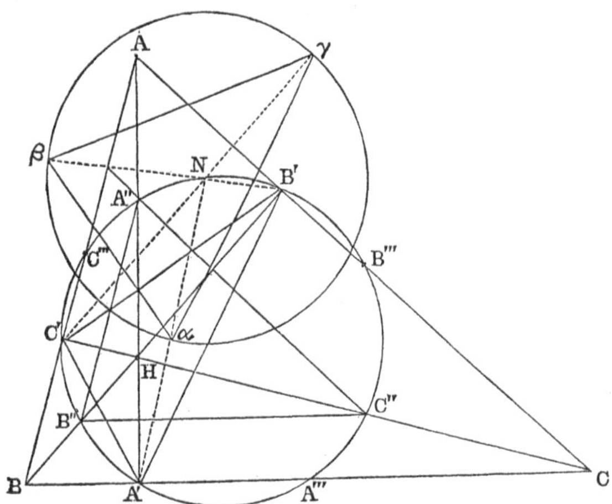

The three angles α_1, α_2, α_3 are $\pi - A$, $\pi - B$, $\pi - C$.

First triad of corresponding lines; perpendiculars at the middle points of the corresponding lines AB′, A′B, A′B′.

Second triad of corresponding lines; perpendiculars at the middle points of the corresponding lines B′C′, BC′, B′C.

Third triad of corresponding lines; perpendiculars at the middle points of the corresponding lines C′A, C′A′, CA′.

The point of concurrence of these triads are the middlo points A‴, B‴, C‴ of the sides of ABC.

The point of concurrence of the lines of F_1 of these triads is the middle A″ of AH; the point of concurrence of the lines of F_2 is the middle B″ of BH; and of the lines of F_3 the middle C″ of CH.

The points A″, B″, C″ are the invariable points. Hence the nine points, viz., A′, B′, C′ (centres of similitude); A″, B″, C″ (invariable points); A‴, B‴, C‴ (points of concurrence of triads of corresponding lines), are on the circle of similitude. Hence the circle of similitude is the nine-points circle of the triangle.

Hence we have the following theorems:—

1°. *Three homologous lines of the triangles* AB′C′, A′BC′, A′B′C *form a triangle* $\alpha\beta\gamma$ *in perspective with* A′B′C′; *the centre of perspective,* N, *is on the nine-points circle of* ABC, *and it is the circumcentre of* $\alpha\beta\gamma$. For its distances to the sides of $\alpha\beta\gamma$ are : : cos A : cos B : cos C. For example, the Brocard lines of the three triangles possess this property.

2°. *Lines joining the points* A″, B″, C″ *to three homologous points* F_1, F_2, F_3 *are concurrent, and meet on the nine-points circle of* ABC.

3°. *If* P, P_1, P_2, P_3 *be corresponding points of the triangles* ABC, AB′C′, A′BC′, A′B′C, *the lines* A″P_1, B″P_2, C″P_3 *meet the nine-points circle of* ABC *in the point which is the isogonal conjugate with respect to the triangle* A″B″C″ *of the point of infinity on the line joining* P *to the circumcentre of* ABC.

4°. *Every line passing through the orthocentre* H *meets the circumcircles of the triangles* AB′C′, A′BC′, A′B′C *in corresponding points.*

5°. *The lines joining the points* A″, B″, C″ *to the centres of the inscribed circles of the triangles* AB′C′,

A′BC′, A′B′C, *pass through the point of contact of the nine-points circle of* ABC *with its inscribed circle.*

Dem.—Let O be the circumcentre. Join OA, and draw A″E parallel to OA, meeting OH in E; then EA′ is a radius of the nine-points circle. Let AD be the bisector of the angle BAC; then the incentres of the

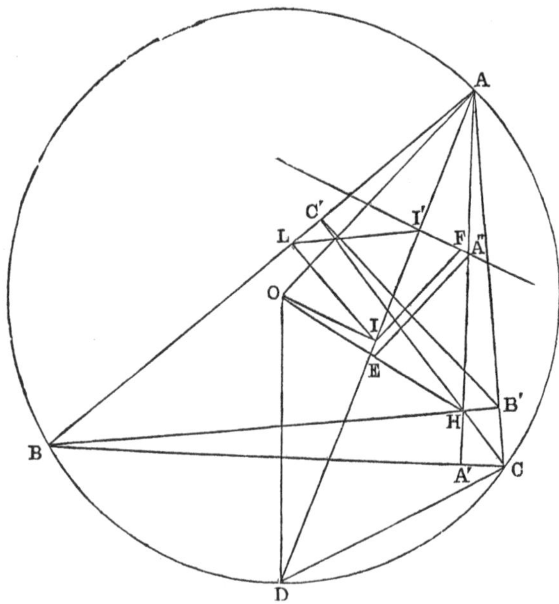

triangles ABC, ABC′ are in the line AD. Let these be I, I′. Join I′A″. It is required to prove that I′A″ passes through the point of contact of the nine-points circle with the incircle of ABC. From I let fall the perpendicular IL on AB. Join LI′. It is easy to see that the triangle ILI′ is isosceles;—in fact IL is equal to LI′. Hence if r be the inradius of the triangle ABC,

R the circumradius, we have $2r^2 = AI \cdot II'$, and $2Rr = AI \cdot ID$. Hence $r : R :: II' : DI$.

Again, through I draw IF parallel to EA″. Now, since the points I′, A″, in the triangle AB′C′, correspond to I and O in ABC, the angle AI′A″ = AIO. Hence the angle II′F is equal to DIO, and the angle I′IF is equal to IDO, because each is equal to DAO. Hence the triangles II′F and DIO are equiangular. Therefore $II' : DI :: IF : DO$. Hence $IF : DO :: r : R$. Therefore $IF = r$. Now since EA″ and IF are parallel, and are radii respectively of the nine-points circle, and incircle of ABC, the line FA″ passes through their centre of similitude. *Hence the proposition is proved.*

Similarly, *if* J′ *be the centre of any of the escribed circles of the triangle* AB′C′, *the line* A″J′ *passes through the point of contact of the nine-points circle of* ABC *with the corresponding escribed circle.*

Exercises.

1. If A_1, B_1, C_1 be the reflexions of the angular points A, B, C of the triangle ABC, with respect to the opposite sides, then the triangles A_1BC, AB_1C, ABC_1, being considered as portions of three figures directly similar,

Prove that—

(1°) A, B, C are the double points.

(2°) The orthocentres of A_1BC, AB_1C, ABC_1, are the invariable points.

(3°) A_1, B_1, C_1 are the adjoint points.

(4°) The orthocentre of ABC is the director point.

(5°) The incentre of the triangle formed by three homologous lines is its perspective centre.

(6°) The triangle formed by any three homologous lines is similar to the orthocentric triangle of ABC.

(7°) The lines joining the orthocentres of A_1BC, AB_1C, ABC_1 to their incentres are concurrent.

2. If through the orthocentre of a triangle be drawn any line L meeting the sides in A′, B′, C′, the lines through A′, B′, C′, which are the reflections of L, with respect to the sides of the triangle, are concurrent.

3. If upon the sides of a triangle, ABC, be constructed three triangles, BCA_1, CAB_1, ABC_1, such that A is an excentre of BCA_1, B an excentre of CAB_1, C an excentre of ABC_1,

Prove that—

(1°) The triangles BCA_1, CAB_1, ABC_1, are directly similar.

(2°) A, B, C are the double points.

(3°) The incentres of A_1BC, B_1CA, C_1AB are the invariable points.

(4°) A_1, B_1, C_1 are the adjoint points.

(5°) The circumcentre of ABC is the director point.

(6) The perspective centre of the triangle formed by three homologous lines is the orthocentre of that triangle.

(7°) Three homologous lines form a triangle inversely similar to ABC.

(8°) The lines joining the incentre of A_1BC, AB_1C, ABC_1, are concurrent.

4. Prove that the triangles, AE′F, DBF′, D′EC, fig., p. 183, are directly similar, and that—

(1°) The invariable points are the centroids of these triangles.

(2°) The double points are the intersections of the symmedians of the triangle ABC with the circle through the invariable points.

(3°) The director point is the symmedian point of ABC.

(4°) The perspective centre of the triangle formed by any three homologous lines is the centroid of that triangle.

The application of Tarry's theory of similar figures contained in this sub-section, with the exception of the theorems 3° and 5°, and the Exercises 2 and 4, are due to Neuberg. The demonstration of 5°, given in the text, is nearly the same as one given by Mr. M'Cay shortly after I communicated the theorem to him.

SECTION VI.

Theory of Harmonic Polygons.

Def. i.—*A cyclic polygon of any number of sides, having a point* K *in its plane, such that perpendiculars from it on the sides are proportional to the sides, is called a harmonic polygon.*

Def. ii.—*The point* K *is called the symmedian point of the polygon.*

Def. iii.—*The lines drawn from* K *to the angular points of the polygon are called its symmedian lines.*

Def. iv.—*Two figures having the same symmedian lines are called co-symmedian figures.*

Def. v.—*If* O *be the circumcentre of the polygon, the circle on* OK, *as diameter, is called its Brocard circle.*

Def. vi.—*If the sides of the polygon be denoted by* $a, b, c, d, \ldots$ *and the perpendiculars on them from* K *by* $x, y, z, u, \ldots$ *then the angle* ω, *determined by any of the equations* $x = \frac{1}{2} a \tan \omega$, $y = \frac{1}{2} b \tan \omega$, &c., *is called the Brocard angle of the polygon.*

Prop. 1.—*The inverses of the angular points of a regular polygon of any number of sides, with respect to any arbitrary point, form the angular points of a harmonic polygon of the same number of sides.*

Dem.—Let A, B, C . . . be the angular points of the regular polygon; A′, B′, C′ . . . the points diametrically opposite to them. Now, inverting from any arbitrary point, the circumcircle of the regular polygon will invert into a circle X, and its diameters AA′, BB′, CC′ . . . into a coaxal system Y, Y_1, Y_2, &c; then [VI., Section v., Prop. 4] the radical axes of the pairs of circles X, Y; X, Y_1; X, Y_2, &c., are concurrent.

Hence, if the inverses of the systems of points A, B, C . . . A′, B′, C′ . . . be the systems $\alpha\beta\gamma$, . . . $\alpha'\beta'\gamma'$ . . ., the lines $\alpha\alpha'$, $\beta\beta'$, $\gamma\gamma'$ . . . are concurrent. Let their common point be K; and since, evidently, the points A, B, C, B′ form a harmonic system, their inverses, the points α, β, γ, β', form a harmonic system; but the line $\beta\beta'$ passes through K. Therefore the perpendiculars from K on the lines $\alpha\beta$, $\beta\gamma$ are proportional to these lines. Hence the proposition is proved.

Cor. 1.—*If the vertices of a harmonic polygon of n sides be* 1, 2, 3 . . . *n, and* K *its symmedian point, the re-entrant polygon formed by the chords* $\overline{13}$, $\overline{24}$, $\overline{35}$, &c., *is a harmonic polygon, and* K *is its symmedian point.*

This is proved by showing that the perpendiculars from K on these chords are proportional to the chords.

Thus, let A, B, C be any three consecutive vertices; p, p' perpendiculars from K on the lines AB, AC; and let AK produced meet the circumcircle again in A; then it is easy to see that the ratio $\frac{p}{AB} : \frac{p'}{AC}$ is equal to the anharmonic ratio (ABCA′), which is constant, because [Book VI., Section iv., Prop. 9] it is equal to the corresponding anharmonic ratio in a regular polygon, and $\frac{p}{AB}$ is constant. Hence $\frac{p'}{AC}$ is constant.

Cor. 2.—*In the same manner the polygon formed by the chords* 14, 25, 36 *is a harmonic polygon, and K is its symmedian point,* &c.

Cor. 3.—*The vertices of any triangle may be considered as the inverses of the angular points of an equilateral triangle.*

Cor. 4.—*A harmonic quadrilateral is the inverse of a square; and its symmedian point is the intersection of its diagonals.*

Cor. 5.—*A harmonic quadrilateral is the figure whose vertices are four harmonic points on a circle* [Book VI., Sect. III., Prop. 9, Cor. 2]. *Hence, the rectangle contained by one pair of opposite sides is equal to the rectangle contained by the other pair.*]

Cor. 6. *If* 1, 2, 3 . . . $2n$ *be the vertices of a harmonic polygon of an even number of sides, the polygon formed by the alternate vertices* 1, 3, 5 . . . $2n - 1$ *is a harmonic polygon, and so is the polygon formed by the vertices* 2, 4, 6, . . . $2n$, *and these three polygons have a common symmedian point.*

Prop 2.—*To invert a harmonic polygon into a regular polygon.*

Sol.—Let AB be a side of the harmonic polygon, Z its circumcircle, O the circumcentre, and K the symmedian point.

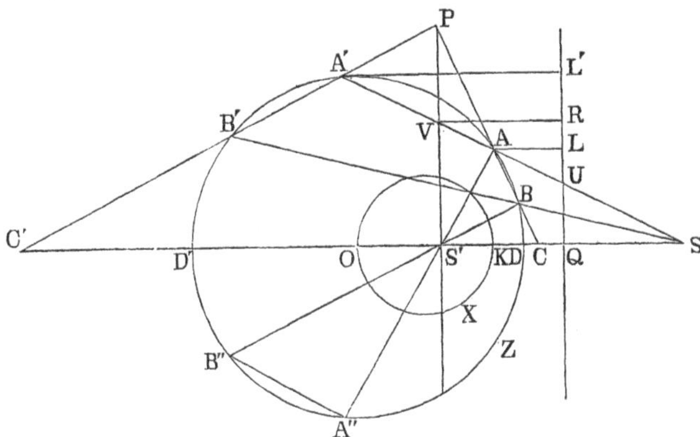

Upon OK as diameter describe a circle OKX; and let S, S′ be the limiting points of Z and OKX; join SA, SB, and produce, if necessary, to meet Z again in A′, B′. Then A′B′ is the side of a regular polygon.

Dem.—Let AB, A′B′, produced, intersect in P, and meet the line OK in C, C′. Now the polar of S will pass through P and through S′; then the pencil P(SCS′C′) is harmonic.

$$\therefore\ 2/SS' = 1/SC + 1/SC' = 1/SK + 1/SO.$$

Hence $(SK - SC) / (SK . SC) = (SC' - SO) / (SC' . SO)$;

$$\therefore\ KC/SC : OC'/SC :: SK : SO;$$

or, $(K, AB) / (S, AB) : (O, A'B') / (S, A'B') :: SK : SO$.

But $(S, AB) : (S, A'B') :: AB : A'B'$. [Book VI., Sect. IV., Prop. 6.]

Hence $(K, AB) / AB : (O, A'B') / A'B' :: SK : SO$.

Now, since AB is a side of a harmonic polygon whose symmedian point is K, the ratio (K, AB) / AB is constant; and since S, K, O are given points, the ratio SK : SO is given; hence the ratio (O, A′B′) / A′B′ is constant; ∴ A′B′ is constant. Hence the proposition is proved.

Cor. 1.—*If we join the points* A, B *to* S′, *and produce to meet* Z *again in* A″, B″, *the points* A″, B″ *are the reflexions* of A, B, with respect to the diameter DD′ of Z.

Def. vii.—*The points* S, S′ *are called the centres of inversion of the harmonic polygon.*

Cor. 2.—*The centres of inversion of a harmonic polygon are harmonic conjugates with respect to its circumcentre and symmedian point.*

Observation.—It is evident that this proposition gives a new demonstration of Prop. I. It is also plain, if, instead of O, we take a point K′ collinear with O and K, and repeat the foregoing construction, only using K′ instead of O, that we shall have the harmonic polygon, whose symmedian point is K, inverted into another whose symmedian point is K′.

Prop. 3.—*If* δ *be the distance of the symmedian point* K *from the circumcentre, and* R *the circumradius of the polygon*, $\tan\omega = \sqrt{(1 - \delta^2/R^2)} \cot \pi/n$.

Dem.—We have Def. VI. $(K, AB)/AB = 2 \tan\omega$; and since the polygon whose side is $A'B'$ is regular, and has n sides $(O, A'B')/A'B' = 2 \cot \pi/n$. Hence $\tan\omega : \cot \pi/n :: SK : SO$. Again, since the points O, K are harmonic conjugates to S, S′, and S, S′ are inverse points with respect to Z, it is easy to see that

$$SK/SO = \sqrt{(1 - \delta^2/R^2)}.$$

Hence, $$\tan\omega = \sqrt{(1 - \delta^2/R^2)} \cot \pi/n.$$

Cor. 1.—$\delta^2 = R^2 (1 - \tan^2\omega \tan^2 \pi/n)$.

Cor. 2.—*If two harmonic polygons of* m *and* n *sides respectively have a common circumcircle and symmedian point, the tangents of their Brocard angles are* $:: \cot \pi/m : \cot \pi/n$.

Cor. 3.—Since the side $A'B'$ of a regular polygon of n sides may have any arbitrary position as a chord in the circle, *it follows that an indefinite number of harmonic polygons of* n *sides, and having a common symmedian point, can be inscribed in the circle.*

Cor. 4.—*The anharmonic ratio of any four consecutive vertices of a harmonic polygon is constant.*

Prop. 4.—If A_0, $A_1 \ldots A_{n-1}$ be the vertices of a harmonic polygon of n sides, the chords A_1A_{n-1}, A_2A_{n-2}, are concurrent.

Dem.—Let K be the symmedian point. Join A_0K, and produce to meet the circumcircle again in A_0'. Then the points A_0, A_0' are harmonic conjugates with respect to the points A_1, A_{n-1} (Demonstration of Prop. 1).

Hence the line A_1A_{n-1} passes through α, the pole of A_0A_0'. Similarly, A_2A_{n-2} passes through α, &c. Hence the proposition is proved.

COR.—*The vertices* A_1A_{n-1}, A_2A_{n-2}, *&c., form a system of points in involution, and points* A_0, A_0' *are the double points. Hence the vertices of a harmonic polygon form as many systems in involution as it has symmedian lines.*

Prop. 5.—*If a transversal through the symmedian point* K *cuts the sides of the polygon in the points* R_1, R_2 ... R_n, *and if a point* P *be taken on it so that* $1/KR_1 + 1/KR_2 \ldots = n/KP$, *the locus of* P *is the polar of* K *with respect to the circumcircle.*

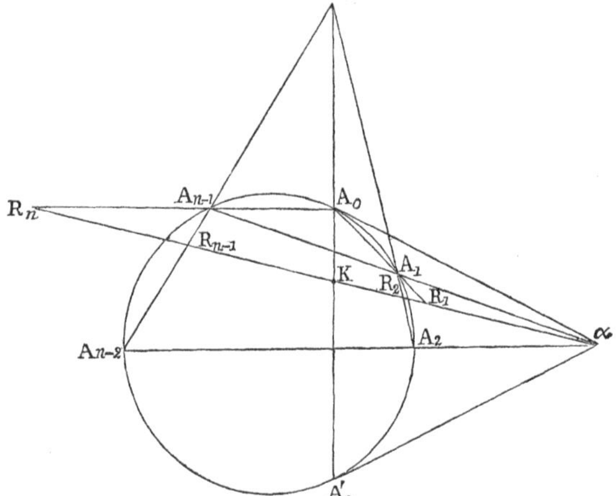

Dem.—Let α be the pole of the symmedian chord A_0A_0'. Join $K\alpha$. It is easy to see that α is one position of P. For if n be even, the sides may be distributed in pairs, so that the points K, α are harmonic conjugates to the points in which each pair of sides may be cut by the line $K\alpha$. Hence,

$$1/KR_1 + 1/KR_n = 2/K\alpha,$$
$$1/KR_2 + 1/KR_{n-1} = 2/K\alpha, \quad \&c.$$

If n be odd, the intercept made by one of the sides on $K\alpha$ is equal to $K\alpha$. Hence, in each case, the sum of the reciprocals of the intercepts made by all the sides on $K\alpha$ is equal to $n/K\alpha$. Therefore α is a point on the locus of P. Similarly, the pole of each symmedian line is a point on the locus. Hence the locus passes through the poles of all the symmedian lines; and since it must be a right line [Ex. 62, page 155], it is the right line through these points. Hence the proposition is proved.

Cor. 1.—*The point* K *and its polar, with respect to the circumcircle, are harmonic pole and polar with respect to the polygon.* (See Salmon's *Higher Curves*, Third Edition, p. 115.)

The harmonic pole and polar are called by French geometers *The Lemoine point and line of the polygon.*

Cor. 2.—*If a harmonic polygon be reciprocated with respect to its Lemoine point, the pole of its Lemoine line is the mean centre of the vertices of the reciprocal polygon.*

This follows from Prop. 5 by reciprocation.

Prop. 6.—*If the lengths of the sides of a harmonic polygon be a, b, c, &c., and the perpendiculars on them from any point* P *in the Lemoine line be α, β, γ, &c., then the sum* $\alpha/a + \beta/b + \gamma/c +$ &c., $= 0$.

Dem.—Let KP intersect the sides in the points R_1, R_2, &c. Then we have

$$(1/KR_1-1/KP)+(1/KR_2-1/KP)+\ldots(1/KR_n-1/KP)=0.$$

Hence

$$PR_1/KR_1 + PR_2/KR_2 + \ldots PR_n/KR_n = 0.$$

Now, if the perpendiculars from K on the sides of the polygon be α', β', γ', &c., $PR_1/KR_1 = \alpha/\alpha'$, $PR_2/KR_2 = \beta/\beta'$, &c. Hence $\alpha/\alpha' + \beta/\beta' + \gamma/\gamma'$, &c., $= 0$; but α', β', γ', &c., are proportional to a, b, c, &c. Hence the proposition is proved.

Prop. 7.—*If the perpendiculars from the vertices of a harmonic polygon on its Lemoine line be denoted by* p_1, p_2 . . . p_n, *and the perpendiculars on it from the Lemoine point by* π, then $\Sigma\,(1/p) = n/\pi$.

Dem.—Let LL′ (fig., Prop. 2) be the Lemoine line. Then, since the points O, S′, K, S form a harmonic system R^2/OS', R^2/OK', R^2/OS are in AP [Book VI., Sect. III., *Cor.*]; that is, OS, OQ, OS are in AP. Therefore S′S is bisected in Q. Hence VS is bisected in V; and since the points A′, A are harmonic conjugates to V, S, the lines UA, UV, UA′ are in GP [Book VI., Sect. III., Prop. 1]. Therefore AL, VR, A′L′ are in GP. Hence $AL \cdot A'L' = VR^2 = S'Q^2 = OQ \cdot KQ$; that is, $p_1 \cdot A'L' = \pi \cdot OQ$. Therefore $A'L'/\pi = OQ/p_1$. Hence $\Sigma(A'L/\pi) = OQ\,\Sigma(1/p)$. But since A′ is the vertex of a regular polygon whose centre is O, $\Sigma(A'L') = nOQ$. Hence $n/\pi = \Sigma(1/p)$.

Prop. 8.—*If a transversal through the symmedian point* (K) *meet the sides of a harmonic polygon of n sides in the points* R_1, R_2, . . . R_n, *and meet the circumcircle in* P; *then* $1/R_1P + 1/R_2P$ . . . $1/R_nP = n/KP$.

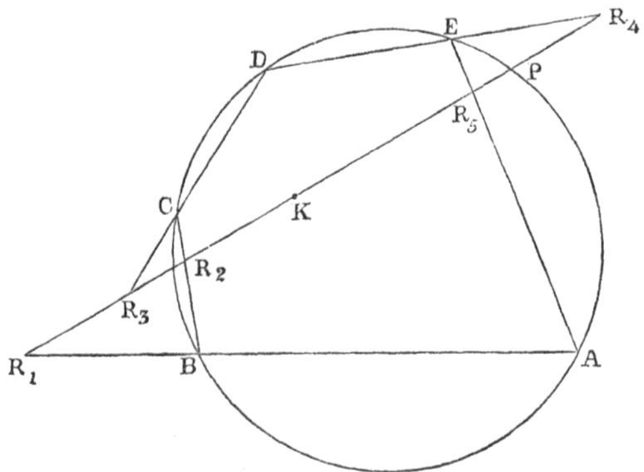

Dem.—Let *a*, *b*, *c*, &c., be the lengths of the sides; α, β, γ, &c., the perpendiculars on them from the point

P; and α', β', γ', &c., the perpendiculars from K. Then [Book VI., Sect. IV., Prop. 7] $a/\alpha + b/\beta + c/\gamma +$ &c., $= 0$; but since K is the symmedian point, α', β', γ', &c., are proportional to a, b, c, &c. Hence $\alpha'/\alpha + \beta'/\beta + \gamma'/\gamma$, &c., $= 0$. Now, $\alpha'/\alpha = R_1K/R_1P = 1 - KP/R_1P$, with similar values for β'/β, &c. Hence,

$$n - KP/R_1P - KP/R_2P \ldots - KP/R_nP = 0;$$

therefore

$$1/R_1P + 1/R_2P \ldots 1/R_nP = n/KP.$$

Prop. 9.—*If through the symmedian point of a harmonic polygon a parallel be drawn to the tangent at any of its vertices, the intercept on the parallel between the symmedian point and the point where it meets either of the sides of the polygon passing through that vertex is constant.*

Dem.—Let AB be a side of the harmonic polygon, AT the tangent, KU the parallel. Produce AK to meet the circle in A′. Join A′B, and let KX be the perpendicular from K on AB; then KX ÷ KU = sin AUK = sin UAT = sin AA′B = $\frac{1}{2}$AB ÷ R. Hence KU ÷ R = KX ÷ $\frac{1}{2}$AB = tan ω (if ω be the Brocard angle of the polygon). Hence KU = R tan ω.

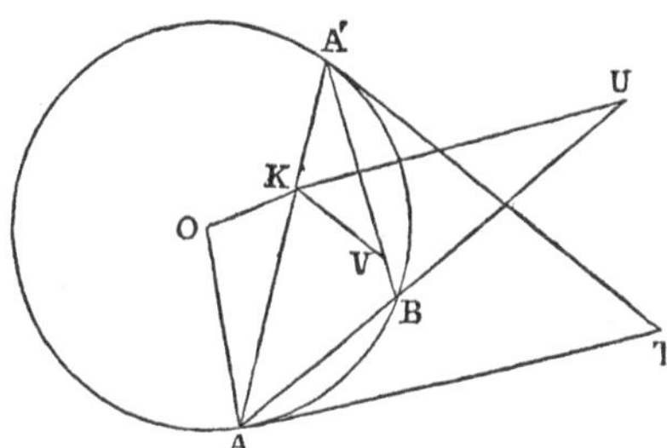

Cor. 1.—*If the polygon consist of n sides there will be 2n points corresponding to* U, *and these points are concyclic.*

Cor. 2.—*If* ω' *be the Brocard angle of the harmonic polygon of which* A′B *is a side*, $\tan\omega \cdot \tan\omega' = \dfrac{\text{AK}.\text{KA}'}{\text{R}^2}$.

For, draw KV parallel to A′T. It is easy to see that the triangles AKU, VKA′ are equiangular. Hence KU . KV = AK . KA′—that is, $\text{R}^2 \tan\omega \cdot \tan\omega' = \text{AK} \cdot \text{KA}'$.

Prop. 10.—*If all the symmedian lines,* KA, KB, &c., *of a harmonic polygon be divided in the points* A″, B″, &c., *in a given ratio, and through these points parallels*

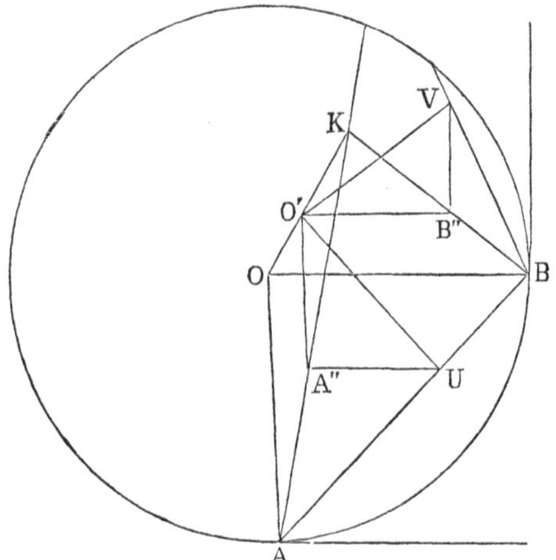

be drawn to the tangents at the vertices, each parallel meeting the two sides passing through the corresponding vertex, all the points of intersection are concyclic, and, taken alternately, they form the vertices of two harmonic polygons.

Dem.—Let KA be divided in A″ in the ratio $l : m$. Join AO, OK, and draw A″O′ parallel to AO, and let A″U be parallel to the tangent AT. Then we have $O'U^2 = O'A''^2 + A''U^2$; but $O'A'' = \dfrac{lR}{l+m}$, and $A''U = KU \dfrac{m}{l+m} = R \tan \omega \dfrac{m}{l+m}$. Hence $(l+m)^2\, O'U^2 = R^2 (l^2 + m^2 \tan^2\omega)$. Hence O′U is constant.

Again, if B″V be parallel to the tangent at B, the triangle O′B″V is in every respect equal to O′A″U. Hence the angle UO′V is equal to AOB. Therefore the points U, V . . . are the angular points of a polygon similar to that formed by the points A, B . . . Hence they form a harmonic polygon. It is evident, by proceeding in the opposite direction from A, that we get another harmonic polygon. Hence the proposition is proved.

Cor. 1.—*The intercept which the circle* O′ *makes on the side* AB *is* $2R\,(l \sin A - m \cos A \tan \omega)/(l + m)$, *where* A *denotes the angle of intersection of the side* AB *with the circumcircle.*

For the perpendiculars from O and K on AB are, respectively,

$$R \cos A, \quad R \sin A \tan \omega,$$

and OK is divided in O′ in the ratio $m : l$. Hence the perpendicular from O′ on AB is

$$R\,(l \cos A + m \sin A \tan \omega)/(l + m)\,;$$

and subtracting the square of this from the square of the radius of O′ we get

$$R^2\,(l \sin a - m \cos A \tan \omega)^2/(l + m)^2.$$

Hence intercept $= 2R\,(l \sin A - m \cos A \tan \omega)/(l + m)$.

Cor. 2.—*By giving special values to the ratio* $l : m$, *we get some interesting results. Thus—*

1°. If $l = 0$ we get intercepts proportional to $\cos A$,

cos B, cos C, &c. *The circle in this case is that of* Prop. 9, *and is called the* COSINE CIRCLE *of the polygon.*

2°. If $l = m$, the line OK will be bisected in O′, and the circle will be concentric with the Brocard circle. *This is, by analogy, called the* LEMOINE CIRCLE *of the polygon: its radius is equal to* R sec ω, *and the intercepts which it makes on the sides are proportional to* $\sin(A - \omega)$, $\sin(B - \omega)$, *&c.*

3°. If $l = m \tan \omega$, *the intercepts are proportional to* $\sin(A - \pi/4)$, $\sin(B - \pi/4)$, *&c., and the radius is equal* $R \sin \omega \operatorname{cosec}(\omega + \pi/4)$.

4°. If $l = m \tan^2 \omega$. *The centre of the circle is the middle point of the line* $\Omega\Omega'$, *its radius is equal to* $R \sin \omega$, *and the intercepts are proportional to* $\cos(A+\omega)$, $\cos(B+\omega)$, *&c. These will be the projections of* $\Omega\Omega'$ *on the sides of the polygon.*

5°. If the polygon reduce to a triangle, and the ratio of $l : m$ be

$$- \cos A \cos B \cos C : 1 + \cos A \cos B \cos C,$$

the intercepts are, respectively, equal to

$$R \sin 2A \cos(B - C),$$
$$R \sin 2B \cos(C - A),$$
$$R \sin 2C \cos(A - B).$$

The perpendiculars from the centre on the sides will be proportional to $\cos^2 A$, $\cos^2 B$, $\cos^2 C$. *This is the case of Taylor's Circle. The ratio* $l : m$ *expressed in terms of* ω *is* $\sin(A - \omega) - \sin^3 A : \sin^3 A$.

6°. *Any Tucker's circle of the triangle* ABC *is a Taylor's circle of some other triangle having the same circumcircle and symmedian point.*

For the Tucker's circle of the triangle ABC being given, the ratio $l : m$ is given, and from the proportion $\sin(A - \omega) - \sin^3 A : \sin^3 A :: l : m$, we get, putting $\cot A = x$, the equation

$$x^3 - \cot \omega . x^2 + x + (1 + l/m) \operatorname{cosec} \omega - \cot \omega = 0;$$

the three roots of which are the cotangents of the three angles of the required triangle.

11. *If* ABC . . ., A′B′C′, . . . *be two homothetic harmonic polygons of any number of sides;* K *their homothetic centre; and if consecutive pairs of the sides of* A′B′C′ . . . *produced, if necessary, intersect the corresponding pairs of* ABC . . . *in the pairs of points* $\alpha\alpha'$, $\beta\beta$, $\gamma\gamma$, &c., *the points* $\alpha\alpha$, $\beta\beta'$, $\gamma\gamma'$, &c., *are concyclic.*

For, since the figure βBβ'B′ is a parallelogram, BB′ is bisected by $\beta\beta'$ in B″; and since the ratio of KB : KB′ is given, the ratio of KB : KB″ is given, and $\beta\beta'$, through B″, is parallel to the tangent at B. Hence, &c.

Cor. 1.—*If the harmonic polygons of Cor.* 2, *be quadrilaterals, their circumcircles and that of the octagon* $\alpha\alpha'\beta\beta'\gamma\gamma'\delta\delta'$ *are coaxal.*

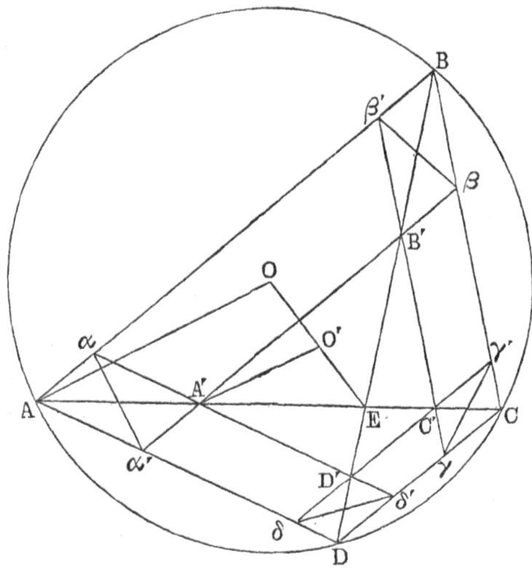

For it is easy to see that the squares of the tangents from B to the circumcircles of $\alpha\alpha'\beta\beta'\gamma\gamma'\delta\delta'$, and A′B′C′D′

are in the ratio 1 : 2; and the squares of tangents from C, D, A to the same circles are in the same ratio.

Cor. 2.—*If the two harmonic polygons of Cor.* 2 *have an even number of sides, the n points of intersection of the sides of the first with the corresponding opposite sides of the second, respectively, are collinear.*

Dem.—For simplicity, suppose the figures are quadrilaterals, but the proof is general. Let P be the point of intersection; then the angle ABE = AC′D′. Hence ABC′D′ is a cyclic quadrilateral. Therefore P is a point on the radical axis of the circumcircles. Hence the proposition is proved.

Cor. 3.—*In the general case the lines $\alpha\alpha'$, $\beta\beta'$, $\gamma\gamma'$ are the sides of a polygon, homothetic with that formed by the tangents at the angular points* A, B, C, &c. *Hence it follows, if the harmonic polygon* ABC . . . *be of an even number of sides, that the intersections of the lines $\alpha\alpha'$, $\beta\beta'$, $\gamma\gamma'$, taken in opposite pairs, are collinear.*

Prop. 12.—*The perpendiculars from the circumcentre of a harmonic polygon, of any number of sides n on the sides, meet its Brocard circle in n points, which connect concurrently in two ways with the vertices of the polygon.*

This general proposition may be proved exactly in the same way as Prop. 2, page 196.

Def.—*If the points of concurrence of the lines in this proposition be* Ω, Ω′, *these are called the* Brocard Points *of the polygon; and the n points* L, M, N, &c., *in which the perpendiculars meet the Brocard circle, for the same reasons as in Cor.* 2, *p.* 194, *are called its* Invariable Points. *Also the points of bisection of the symmedian chords* AK, BK, CK, &c., *will be its* Double Points.

Cor. 1.—*The n lines joining respectively the invariable points* L, M, N . . . *to n corresponding points of figures directly similar described on the sides of the harmonic polygon are concurrent; the locus of their point of concurrence is the Brocard circle of the polygon.*

Prop. 13.—*The centres of similitude of the pairs of consecutive sides of a harmonic polygon form the vertices of a harmonic polygon.*—(Tarry.)

Dem.—Let A be a vertex of the harmonic polygon, K its harmonic pole. Join AK, and produce to meet the Brocard circle in M. Join MS, cutting the Brocard circle in N, and AS, cutting the circumcircle in A′. Join ON. Now the polar PQ of S passes through the

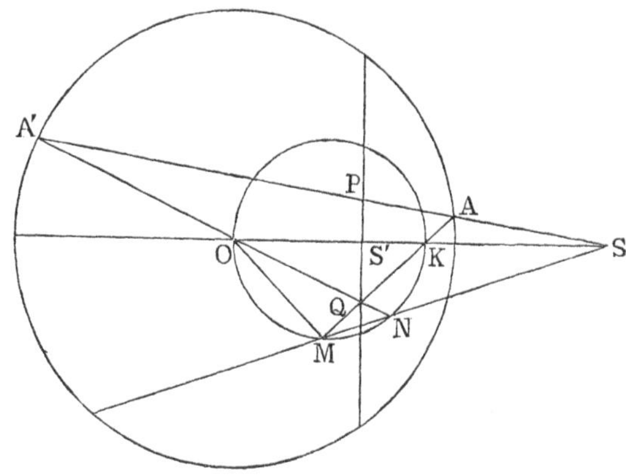

intersection of MK and ON; and since the points P, S are harmonic conjugates to A, A′, and S′, S to K, O, the pencils Q(SKS′O), Q(SAPA′), are equal, and they have three common rays, viz. QS, QA, QP. Hence their fourth rays, QO, QA′, coincide; therefore the points Q, O, A′ are collinear. Again, A′, being the inverse of A with respect to Z, is a vertex of a regular polygon inscribed in Z. Hence N is the vertex of a regular polygon inscribed in the Brocard circle; and therefore M, which is its inverse, is a vertex of a harmonic polygon; but M is evidently the middle point of the symmedian chord passing through A. Hence it is the centre of similitude of the two consecutive sides passing through A. Hence the proposition is proved.

Prop. 14.—*If figures directly similar be described on the sides of a harmonic polygon of any number of sides, the symmedian lines of the harmonic polygon formed by corresponding lines of these figures pass through the middle points of the symmedian chords of the original figures.*

This is an extension of Prop. 5, page 197, and may be proved exactly in the same way.

Cor. 1.—*The symmedian point of the harmonic polygon, formed by corresponding lines of figures directly similar, is a point on the Brocard circle of the original polygon.*

Cor. 2.—*The invariable points of the original polygon are corresponding points of figures directly similar described on its sides.*

Cor. 3.—*The centre of similitude of the original polygon, and that formed by any system of corresponding lines, is a point on the Brocard circle of the original polygon.*

Cor. 4.—*The centre of similitude of any two harmonic polygons, whose sides respectively are two sets of corresponding lines of figures directly similar, described on the sides of the original polygon, is a point on the Brocard circle of the original.*

Exercises.

1. If the symmedian lines through the vertices A, B, C of a triangle meet its circumcircle in the points A′, B′, C′, the triangles ABC, A′B′C′ are cosymmedian.

For since the lines AA′, BB′, CC′ are concurrent, the six points in which they meet the circle are in involution. Hence the anharmonic ratio (BACA′) = (B′A′C′A); but the first ratio is harmonic, therefore the second is harmonic. Hence A′A is a symmedian of the triangle A′B′C′. Similarly, B′B, C′C are symmedians.

2. The centres of inversion of a harmonic polygon are the limiting points of its circumcircle and Brocard circle, and the Lemoine line is their radical axis.

3. The product of any two alternate sides of a harmonic polygon is proportional to the product of the sines of their inclinations to their included side.

For if A, B, C, D be four consecutive vertices, A′, B′, C′, D′ the corresponding vertices of a regular polygon, the anharmonic ratio (ABCD) = (A′B′C′D′). Hence (AB . CD) / (AC . BD) = (A′B′ . C′D′) / (A′C′ . B′D′) = ($\sec^2 \pi/n$) / 4 ; but AC = 2 R sin ABC, BD = 2 R sin BCD.
Hence (AB . CD)/sin ABC . sin BCD = $R^2 \sec^2 \pi/n$.

4. If we invert the sides A′B′ (see fig. p. 207) of a regular polygon with respect to S, we get a circle passing through AB and S. Hence, if through the extremities of each side of a harmonic polygon circles be described passing through either of the centres of inversion, these circles cut the circumcircle at a constant angle π/n.

5. In the same case they all touch another circle, and the points of contact are the vertices of a harmonic polygon.

6. If through the symmedian point, and any two adjacent vertices of a harmonic polygon, a circle be described, it cuts the circumcircle at a constant angle.

7. A system of circles passing through the two centres of inversion of a harmonic polygon, and passing respectively through its vertices, cut each other at equal angles, and cut its circumcircle and Brocard circle orthogonally.

8. In the same case the points of intersection on the Brocard circle are the vertices of a harmonic polygon.

9. Prove that the centre of similitude of the two polygons formed by the alternate vertices in Prop. 9 is the symmedian point of the original polygon, and that the centre of similitude of either, and the original polygon, is a Brocard point of the original polygon.

10. Prove that the circles in Ex. 5, described through the symmedian point, and through adjacent vertices of a harmonic polygon, all touch a circle coaxal with the Brocard circle and the circumcircle, and that the points of contact are the vertices of a harmonic polygon.

11. If A, B, C be any three consecutive vertices of a harmonic polygon, whose symmedian point is K, prove, if K′ be the symmedian point of the triangle ABC, and B′ the point where KK′ intersects AC, that the anharmonic ratio (KB′K′B) is constant.

12. If through the vertices A, B, C, &c., of a harmonic polygon F, be drawn lines making the same angle ϕ with the sides AB, BC, &c., and in the same direction of rotation, prove that the polygon F_2 formed by these lines is a harmonic polygon, and similar to the original.

13. If a polygon F_3 be formed by lines which are the isogonal conjugates of F_2, with respect to the angles of F_1, prove that F_3 is equal to F_2 in every respect.

14. If F_1, F_2, F_3 be considered as three directly similar figures, prove that their symmedian points are the invariable points, and that the double points are the circumcentre of F, and its Brocard points.

15. The symmedian point of a harmonic polygon is the mean centre of the feet of isogonal lines drawn from it to the sides of the polygon.

This follows from the fact that the isogonal lines make equal angles with, and are proportional to, the sides of a closed polygon.

16. The square of any side of a harmonic polygon is proportional to the rectangle contained by the perpendiculars from its extremities on the harmonic polar.

17. If from the angular points of a harmonic polygon tangents $t_1, t_2, \ldots t_n$ be drawn to its Brocard circle, prove that

$$\Sigma(1/t^2) = n/(R^2 - \delta^2).$$

18 If ABCD, &c., be a harmonic polygon, Ω one of its Brocard points, prove that the lines AΩ, BΩ, &c., meet the circumcircle again in points which form the vertices of a harmonic polygon equal in every respect, and that Ω will be one of its Brocard points.

The extension of recent Geometry to a harmonic quadrilateral was made by Mr. Tucker in a Paper read before the Mathematical Society of London, February 12, 1885. His researches were continued by Neuberg in *Mathesis*, vol. v., Sept., Oct., Nov., Dec., 1885. The next generalization was made by me in a Paper read before the Royal Irish Academy, January 26, 1886, "On the Harmonic Hexagon of a Triangle." Both extensions are special cases of the theory contained in this section, the whole of which I discovered since the date of the latter Paper, and which M. Brocard remarks, "parait être le couronnement de ces nouvelles études de géométrie du triangle." The following passage, in a note by Mr. M'Cay in Tucker's Paper, shows the idea of extension had occurred to that geometer:—"I believe all these results would hold for a polygon in a circle, if the side were so related that there existed a point whose distances from the sides were proportional to the sides."—*March*, 1886.

Since the date of the foregoing note, which appeared in the 4th Edition of the "Sequel," two Papers on the Harmonic Polygon have been published; one by MM. Neuberg and Tarry—Congress of Nancy, 1886, *Association Française pour l'avancement des Sciences;* the other by the Rev. T. C. Simmons—*Proceedings of the London Mathematical Society*, April, 1887. I am indebted to the former of these for the demonstration of Prop. 2, and to the latter for the enunciation of Prop. 7. With these exceptions, and Prop. 13, all that is contained in this Section is original.

SECTION VII.

General Theory of Associated Figures.

Def. i.—*If any point* X *in the circumference of a circle* Z *be joined to n fixed points* I_1, I_2, . . . I_n, *on the same circumference, and portions* I_1A_1, I_2A_2 . . . I_nA_n *be taken on the joining lines in given ratios* d_1, d_2 . . . d_n,

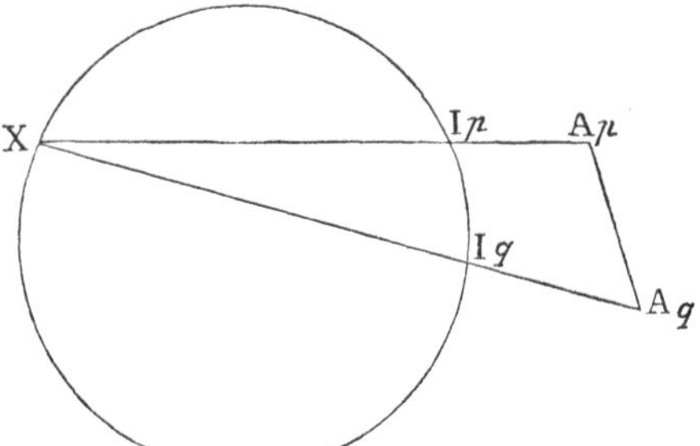

and all measured in the same direction with respect to X, *a system of figures directly similar, described on* I_1A_1, . I_2A_2 . . . I_nA_n, *is called an associated system.*

Def. ii.—*The points* I_1, I_2 . . . I_n *are called the invariable points of the system.*

Prop. 1.—*The centres of similitude of an associated system of figures are concyclic.*

Dem.—Let the figures be $F_1, F_2, \ldots F_n$, and I_p, I_q, any two invariable points, A_p, A_q the corresponding points of F_p, F_q taken on the lines XI_p, XI_q. Hence [Sup. Sect. II., Prop. 4] the second intersection of the circle Z with the circumcircle of the triangle XA_pA_q is the centre of similitude of the figures F_p, F_q. Hence the centre of similitude of each pair of figures of the associated system lies on Z, that is, on the circle through the invariable points.

Def. III.—*The circle Z, through the invariable points, is, on account of the property just proved, called the circle of similitude of the system.*

Prop. 2.—*The figure formed by n homologous points is in perspective with that formed by the invariable points.*

This follows from Def. I.

Cor.—*Every system of n homologous lines passing through the invariable points forms a pencil of concurrent lines.*

Prop. 3.—*In an associated system of n figures the points of intersection of n homologous lines are in perspective with the centres of similitude of the figures.*

Dem.—Let the homologous be $L_1, L_2, \ldots L_n$, and through the invariable points draw lines respectively parallel to them Then, since these parallels are corresponding lines of $F_1, F_2, \ldots F_n$, they are concurrent. Let them meet in K. Now, consider any two lines L_p, L_q : the perpendiculars on them from K are respectively equal to their distances from the invariable points I_p, I_q, and therefore proportional to the perpendiculars on them from the centre of similitude S_{pq} of the figures F_p, F_q. Hence the point of intersection of L_p, L_q, the point K, and S_{pq}, are collinear. Hence it follows that the lines joining the $n(n+1)/2$ intersections of $L_1, L_2, \ldots L_n$ to K pass respectively through the $n\ (n+1)/2$ centres of similitude.

DEF. IV.—*I shall call* K *the perspective centre of the polygon formed by the lines* $L_1, L_2, \ldots L_n$.

Prop 4.—*In an associated system of n figures, the centre of similitude of any two polygons,* G, G′, *each formed by a system of n homologous lines, is a point on the circle of similitude.*

Dem.—Let K, K′ be the perspective centres of G, G′: thus K, K′ are corresponding points of G, G′. Let I be any of the invariable points. Join IK, IK′, and let the joining lines meet any two corresponding lines of G, G′, in N, N′; then KN, K′N′ are corresponding lines of G, G′. Hence the centre of similitude is the second intersection of the circumcircle of the triangle INN′ with the circle of similitude. Hence the proposition is proved.

Prop. 5.—*The six centres of similitude of an associated system of four figures taken in pairs are in involution.*

Dem.—Let L_1, L_2, L_3, L_4 be four homologous lines of the figures, and K the perspective centre of the figure formed by these lines. Then the pencil from K to the six centres of perspective passes [Prop. 3] through the three pairs of opposite intersections of the sides of the quadrilateral L_1, L_2, L_3, L_4, and therefore forms a pencil in involution.

DEF. V.—*If in an associated system of n figures* F_1, $F_2 \ldots F_n$ *there exist* $n+1$ *points* $A_1, A_2 \ldots A_n, A_{n+1}$, *such that* $A_1A_2, A_2A_3, \ldots A_nA_{n+1}$ *are homologous lines of the figures; then the broken line* $A_1A_2 \ldots A_nA_{n+1}$ *is called a* TARRY'S LINE; *and, if* A_{n+1} *coincides with* A_1, *a* TARRY'S POLYGON.

I have named the line of this definition after M. Gaston Tarry, "Receveur des contributions, à Alger," to whom the theory of associated figures is due. See *Mathesis*, Tome, VI., pp. 97, 148, 196, from which this Section, except propositions 4 and 5, is taken.

Prop. 6.—*If* $A_1A_2 \ldots A_nA_{n+1}$ *be a Tarry's line,* $I_1, I_2 \ldots I_n$ *the invariable points, then the angle* $I_1A_1A_2 = I_2A_2A_3 \ldots = I_nA_nA_{n+1}$, *and the angle* $I_1A_2A_1 = I_2A_3A_2 \ldots = I_nA_{n+1}A_n$.

Dem.—By hypothesis the triangles $I_1A_1A_2$, $I_2A_2A_3 \ldots I_nA_nA_{n+1}$ form an associated system. Hence they are equiangular, and therefore the proposition is proved.

Def. VI.—*The lines* A_1I_1, $A_2I_2 \ldots A_nI_n$ *being corresponding lines passing the invariable points, meet on the circle of similitude. In like manner,* A_2I_1, $A_3I_2 \ldots A_{n+1}I_n$ *meet on the circle of similitude. The points of concurrence are called the Brocard points of the system, and denoted by* Ω, Ω'.

Def. VII.—*The base angles of the equiangular triangles* $I_1A_1A_2$, $I_2A_2A_3 \ldots$ *are called its Brocard angles, and denoted by* ω, ω', *viz.* $I_1A_1A_2$ *by* ω, *and* $I_1A_2A_1$ *by* ω'.

Def. VIII.—*The perspective centre* [Def. IV.] *of a Tarry's line, being such that perpendiculars from it on the several parts of that line are proportional to the parts, is called the symmedian point of the line.*

Prop 7.—*Being given two consecutive sides* A_1A_2, A_2A_3 *of a Tarry's line, and its Brocard angles, to construct it.*

Sol.—Upon A_1A_2, A_2A_3 construct two triangles $A_1I_1A_2$, $A_2I_2A_3$, having their base angles equal to ω, ω', respectively, viz., $A_2A_1I_1 = A_3A_2I_2 = \omega$, and $A_1A_2I = A_2A_3I_2 = \omega'$. Then the vertices I_1, I_2 are invariable points. The lines A_1I_1, A_2I_2 will meet in one of the Brocard points Ω [Def. VI.], and the lines A_2I_1, A_3I_2 in the other Brocard point Ω'. Then the four points I_1, I_2, Ω, Ω' are concyclic, and the circle Z, through them, is the circle of similitude. Join ΩA_3. This will meet Z in I_3, which will be the invariable point corresponding to the next side, A_3A_4, of the Tarry's line. Join $\Omega' I_3$, and produce to meet the line A_3A_4, making with ΩA_3 an angle equal to ω in A_4. We construct in the

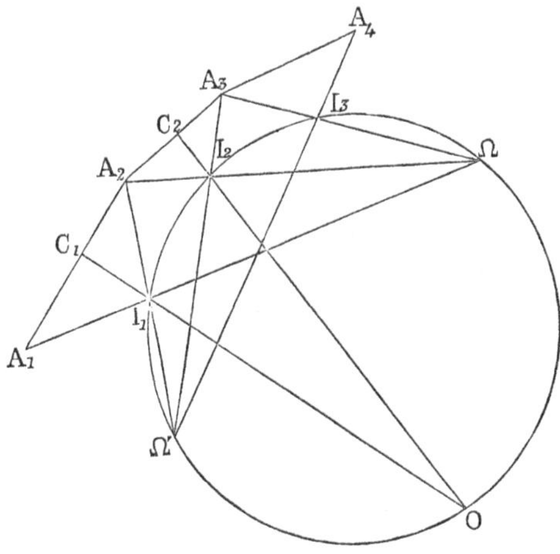

same manner A_4A_5, A_5A_6, &c. Tarry's line, it is obvious, may be continued in the opposite sense, $A_3A_2A_1$.

Cor.—*If one side of a Tarry's line and its Brocard points be given, it may be constructed.*

Prop 8.—*If the Brocard angles* ω, ω' *of a Tarry's line be equal, the points* A_1, A_2 . . . A_n *form the vertices of a harmonic polygon.*

Dem.—From the invariable points I_1, I_2, &c., let fall perpendiculars I_1C_1, I_2C_2, &c., on the lines A_1A_2, A_2A_3, &c. Then, since these perpendiculars are homologous lines of F_1, F_2, they meet on Z. Let O be their point of intersection. Now, since $\omega = \omega'$, the triangle

$I_1A_1A_2$ is isosceles. Hence A_1A_2 is bisected in C_1, and $OA_1 = OA_2$. Similarly $OA_2 = OA_3$, &c. Hence the points $A_1, A_2 \ldots$ are concyclic; and since the polygon formed by them has a symmedian point K, it is a harmonic polygon.

Prop. 9.—*If the Brocard angles ω, ω' of a Tarry's line be unequal, it cannot be a closed polygon.*

Dem.—If ω' be $> \omega$, A_1C_1 is $> C_1A_2$. Hence $OA_1 > OA_2$. Similarly $OA_2 > OA_3$. Hence the points A_1, $A_2 \ldots$ are continually approaching O. Hence the proposition is evident.

Cor.—*When Tarry's line has equal Brocard angles, its symmedian point is diametrically opposite to* O *on the circle of similitude.*

For the parallels to A_1A_2, A_2A_3, &c., through the invariable points, meet in K. Now, since I_1K, I_1O are respectively parallel and perpendicular to A_1A_2, the angle OI_1K is right. Hence OK is the diameter of Z.

Exercises.

1. If an associated system of figures have a common centre of similitude, the figures formed by inverting them from that point form an associated system.

2. In the same case, the figures formed by reciprocating them from the centre of similitude form an associated system.

3. If a series of directly similar triangles be inscribed in a given triangle, they have a common centre of similitude.

4. If a series of directly similar triangles be circumscribed to a given triangle, they have a common centre of similitude.

5. In an associated system of four directly similar figures there exists one system of four homologous points which are collinear.

6. In the same case the four director points of the four triads which are obtained from them are collinear.

7. In three directly similar figures there exists an infinite number of triads of corresponding circles which have the same radical axis.

8. In the same case four systems of three homologous circles can be described to touch a given line.

9. Through any point can be described three homologous circles of the system.

10. Eight systems of three homologous circles can be described to touch a given circle.

11. Every cyclic polygon which has a Brocard point is a harmonic polygon.

12. Every polygonal line which has a symmedian point and a Brocard point, or which has two Brocard points, is a Tarry's line.

13. In every system of three similar figures, F_1, F_2, F_3, there exists an infinite number of three homologous segments, AA′, BB′, CC′, whose extremities are concyclic, and the locus of the centre of the circle X through their extremities is the circle of similitude of F_1, F_2, F_3.

14. The Brianchon's point of the hexagon, formed by the tangent to X at the points A, B, C, A′, B′, C′ is the symmedian of the segments.

15. The projections of the centre O of the circle X on the diagonals of Brianchon's hexagon are the double points of the figures F_1, F_2, F_3.

16. If K be the symmedian point of the segments, and O the centre of X; and if perpendiculars from K on the radii OA, OA′ meet the segment AA′ in the points (a, a'); and if (b, b'), (c, c') be similarly determined on BB′, CC′, the six points are concyclic.

17. The symmedian point of the three segments, aa', bb', cc', is the centre of X.

18. If the lines KA, KA′ be divided in a given ratio, and through the points of division lines be drawn respectively perpendicular to OA, OA′, meeting AA′ in the points (α, α'), the points (α, α'), and the points similarly determined on BB′, CC′, are concyclic.

19. If A, A′ be opposite vertices of a cyclic hexagon; P the pole of the chord AA′; L, L′ the points of intersection of the radii OA, OA′ with the Pascal'sl ine of the hexagon; a, a' the projec-

tions of the centre O on the lines PL, PL′; then (α, α'), and the two other pairs of points similarly determined, are concyclic.

20. Let M, M′ be points of the radii OA, OA′, such that the anharmonic ratios (OAML), (OA′M′L′) are equal. The projections of O on the lines PM, P′M′, and the other pairs of points determined in the same manner from the chords BB′, CC′, are concyclic.

21. The circle of similitude of the three directly similar triangles ABC, FDE, E′F′D′ (fig., p. 177) passes through the Brocard points and Brocard centre of ABC.

22–25. If a, b, c be the points of intersection of the corresponding sides of two equal and directly similar triangles, ABC, A′B′C′, whose centre of similitude is S; then, 1°, if S be the circumcentre of ABC, it is the orthocentre of abc; 2°, if it be the incentre of ABC, it is the circumcentre of abc; 3°, if S be the symmedian point of ABC, it is the centroid of abc; 4°, if it be a Brocard point of ABC, it is a Brocard point of abc.

26. State the corresponding propositions for ABC, and the triangle formed by the lines joining corresponding vertices of ABC, A′B′C′.

27. If a cyclic polygon of an even number of sides ABCD, &c., turn round its circumcentre into the position A′B′C′D′, &c., each pair of opposite sides of the polygon whose vertices are the intersection of corresponding sides are parallel.

28. If a variable chord of a circle divide it homographically, prove that there is a fixed point (Lemoine point) whose distance from the chord is in a constant ratio to its length.

29. In the same case, prove that there are two Brocard points, a Brocard angle, a Brocard circle, and systems of invariable points, and double points.

30. Prove also that the circle can be inverted so that the inverses of the extremities of the homographic chords will be the extremities of a system of equal chords.

Miscellaneous Exercises.

1. If from the symmedian point of any triangle perpendiculars be drawn to its sides, the lines joining their feet are at right angles to the medians.

2. If ACB be any triangle, CL a perpendicular on AB; prove that AC and BL are divided proportionally by the antiparallel to BC through the symmedian point.

3. The middle point of any side of a triangle, and the middle point of the corresponding perpendicular, are collinear with the symmedian point.

4. If K be the symmedian point, and G the centroid of the triangle ABC; then—1°, the diameters of the circumcircles of the triangles AKB, BKC, CKA, are inversely proportional to the medians; 2°, the diameters of the circumcircles of the triangles AGB, BGC, CGA are inversely proportional to AK, BK, CK.

5. *If the base* BC *of a triangle and its Brocard angle be given, the locus of its vertex is a circle.* (Neuberg.)

Let K be its symmedian point. Through K draw FE parallel

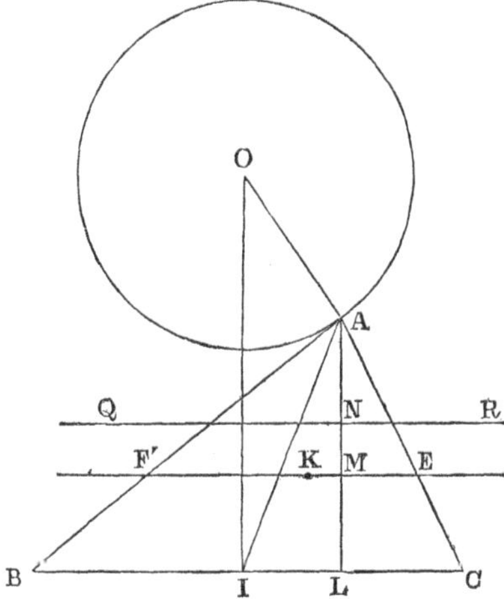

to BC, cutting the perpendicular AL in M. Make MN equal to

half LM. Through L draw QR parallel to PC. Bisect BC in I, and draw IO at right angles, and make $2IO \cdot MN = BI^2$.

Now, because the Brocard angle is given, the line FE parallel to the base through the symmedian point is given in position. Hence QR is given in position; therefore MN is given in magnitude. Hence IO is given in magnitude; therefore O is a given point. Again, because F′E is drawn through the symmedian point, $BA^2 + AC^2 : BC^2 :: AM : ML$; therefore $BI^2 + IA^2 : BI^2 :: MN + NA : MN$. Hence $BI^2 : IA^2 :: MN : NA$; therefore $IA^2 = 2IO \cdot NA$; and since I, O are given points, and QR a given line, the locus of A is a circle coaxal with the point I and the line QR [VI., Section v., Prop. 1].

6. If on a given line BC, and on the same side of it, be described six triangles equiangular to a given triangle, the vertices are concyclic.

7. If from the point I, fig., Ex. 5, tangents be drawn to the Neuberg circle, the intercept between the point of contact and I is bisected by QR.

8. The Neuberg circles of the vertices of triangles having a common base are coaxal.

9. In the fig., Prop. 4, p. 175, if the segment A′B′ slide along the line CB′, prove that the locus of O is a right line.

10. If two triangles be co-symmedians, the sides of one are proportional to the medians of the other.

11. The six vertices of two co-symmedian triangles form the vertices of a harmonic hexagon.

12. The angle BOC, fig., Ex. 5, is equal to twice the Brocard angle of BAC.

13. If the lines joining the vertices of two triangles which have a common centroid be parallel, their axis of perspective passes through the centroid. (M‘Cay.)

14. The Brocard points of one of two co-symmedian triangles are also Brocard points of the other.

15. If L, M, N, P, Q, R be the angles of intersection of the sides of two co-symmedian triangles (omitting the intersections

which are collinear), these angles are respectively equal to those subtended at either Brocard point by the sides of the harmonic hexagon. (Ex. 11.)

16. If two corresponding points, D, E of two directly similar figures, F_1, F_2, be conjugated points with respect to a given circle (X), the locus of each of the points D, E is a circle.

Dem.—Let S be the double point of F_1, F_2, and let DE intersect X in L, M. Bisect DE in N. Join SN. Then, from the property of double points, the triangle SDE is given in species; therefore the ratio SN : ND is given. Again, because E, D (hyp.) are harmonic conjugates with respect to L, M, and N is the middle point of ED, ND^2 is equal to the rectangle NM . NL; that is, equal to the square of the tangent from N to the circle X. Hence the ratio of SN to the tangent from N to X is given. Hence the locus of N is a circle, and the triangle SND is given in species; therefore the locus of D is a circle.

17. If we consider each side of a triangle ABC in succession as given in magnitude, and also the Brocard angle of the triangle, the triangle formed by the centres of the three corresponding Neuberg's circles is in perspective with ABC.

18. If in any triangle ABC triangles similar to its co-symmedian be inscribed, the centre of similitude of the inscribed triangles is the symmedian point of the original triangle.

19. If figures directly similar be described on the sides of a harmonic hexagon, the middle point of each of its symmedian lines is a double point for three pairs of figures.

20. If F_1, F_2, F_3, F_4 be figures directly similar described on the sides of a harmonic quadrilateral, K its symmedian point, K', K'' the extremities of its third diagonal, and if the lines KK', KK'' meet the Brocard circle again in the points H, I; H is the double point of the figures F_1, F_3; I of the figures F_2, F_4.

DEF.—*The quadrilateral formed by the four invariable points of a hormonic quadrilateral is called Brocard's first quadrilateral, and that formed by the middle points of its diagonals, and the double points* H, I, *Brocard's second quadrilateral.*

This nomenclature may evidently be extended.

21. Brocard's second quadrilateral is a harmonic quadrilateral.

22. If ω be the Brocard angle of a harmonic quadrilateral ABCD, $\operatorname{cosec}^2 \omega = \operatorname{cosec}^2 A + \operatorname{cosec}^2 B = \operatorname{cosec}^2 C + \operatorname{cosec}^2 D$.

23. If the middle point F of the diagonal AC of a harmonic quadrilateral be joined to the intersection K′ of the opposite sides AB, CD, the angle AFK′ is equal to the Brocard angle. (Neuberg.)

24. The line joining the middle point of any side of a harmonic quadrilateral to the middle point of the perpendicular on that side, from the point of intersection of its adjacent sides, passes through its symmedian point.

25. If F_1, F_2, F_3 . . . be figures directly similar described on the sides of a harmonic polygon ABC . . . of any number of sides, and if $\alpha\beta\gamma$ . . . be corresponding lines of these figures; then if any three of the lines $\alpha\beta\gamma$ . . . be concurrent, they are all concurrent.

26. In the same case, if the figure ABC . . . be of an even number of sides, the middle points of the symmedian chords of the harmonic polygon $\alpha\beta\gamma$ . . . coincide with the middle points of the symedian chords of ABC.

27. If F_1, F_2, F_3 be three figures directly similar, and B_1, B_2, B_3 three corresponding points of these figures; then if the ratio of two of the sides $B_1B_2 : B_2B_3$ of the triangle formed by these points be given, the locus of each is a circle; and if the ratio be varied, the circles form two coaxal systems.

Dem.—Let S_1, S_2, S_3 be the double points: then the triangles $S_3B_1B_2$, $S_1B_2B_3$ are given in species. Hence the ratios $B_1B_2 : S_3B_2$, and $B_2B_3 : S_1B_2$ are given; and the ratio $B_1B_2 : B_2B_3$ is given by hypothesis. Hence the ratio $S_3B_2 : S_1B_2$ is given, and therefore the locus of B_2 is a circle.

28. If through the symmedian point K of a harmonic polygon of n sides be drawn a parallel to any side of the polygon, intersecting the adjacent sides in the points X, X′, and the circumcircle in Y, Y′, then $4XK \cdot KX' \sin^2 \frac{\pi}{n} = YK \cdot KY'$.

29. If the area of the triangle $B_1B_2B_3$, Ex. 27, be given, the locus of each point is a circle.

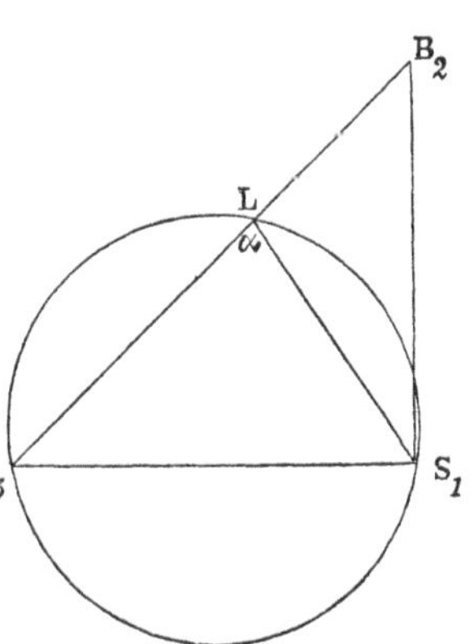

Dem.—Here we have the ratios $B_1B_2 : S_3B_2$, and $B_2B_3 : S_1B_2$ given. Hence the ratio of the rectangle $B_1B_2 . B_2B_3 . \sin B_1B_2B_3 : S_3B_2 . S_1B_2 . \sin B_1B_2B_3$ is given: but the former rectangle is given; therefore the rectangle $S_3B_2 . S_1B_2 . \sin B_1B_2B_3$ is given; now, the angles $B_1B_2S_3$, $S_1B_2B_3$ are given. Hence the angle $B_1B_2B_3 \pm S_3B_2S_1$ is given. Let its value be denoted by α; therefore $B_1B_2B_3 = \alpha \pm S_3B_2S_1$. Hence, taking the upper sign, the problem is reduced to the following. *The base* S_3S_1 *of a triangle* $S_3B_2S_1$ *is given in magnitude and position, and the rectangle* $S_3B_2 . S_1B_2 .$ *sin* $(\alpha - S_3B_2S_1)$ *is given in magnitude, to find the locus of* B_2, which is solved as follows:—Upon S_3S_1 describe a segment of a circle S_3LS_1 [*Euc.* III., XXXIII.] containing an angle S_3LS_1 equal to α. Join S_1L; then the angle B_2S_1L is equal to $\alpha - S_3B_2S_1$. Hence, by hypothesis, $S_3B . S_1B_2 . \sin B_2S_1L$ is given; but $S_1B_2 . \sin B_2S_1L$ is equal to $LB_2 \sin \alpha$; therefore the rectangle $S_3B_2 . LB_2$ is given. Hence the locus of B_2 is a circle.

30. If Q, Q′ be the inverses of the Brocard points of a triangle, with respect to its circumcircle, the pedal triangles of Q, Q′ are—1°, equal to one another; 2°, the sides of one are perpendicular to the corresponding sides of the other; 3°, each is inversely similar to the original. (M'CAY.)

31. If the area of the triangle formed by three corresponding lines of three figures directly similar be given, the envelopes of its sides are circles whose centres are the invariable points of the three figures.

32. If the area of the polygon formed by n corresponding lines of n figures directly similar described on the sides of a harmonic polygon of n sides be given, the envelopes of the sides are circles whose centres are the invariable points of the harmonic polygon.

33. The four symmedian lines of a harmonic octagon form a harmonic pencil.

34, A′, C′ are corresponding points of figures F_2, F_1 directly similar described on the sides BC, AB of a given triangle: if AA′, CC′ be parallel, the loci of the points A, C are circles.

Dem.—Let S be the double point of the figures, and D the point of F_2, which corresponds to C in F_1. Join DA′, and produce CC′ to meet it in E. Now, since S is the double point of the figures ABCC′, BCDA′, the triangles SCC′, SDA′ are equiangular; therefore the angle SCC′ is equal to SDA′. Hence the points S, D, E, C, are concyclic; therefore the angle DEC is equal to the supplement of DSC; that is, equal to ABC; therefore DA′A is equal to ABC, and is given. Hence the locus of A′ is a circle.

35. The Brocard angles of the triangles ABC, BCA′ are equal.

Dem.—Let the circle about the triangle DA′A cut AB in F.

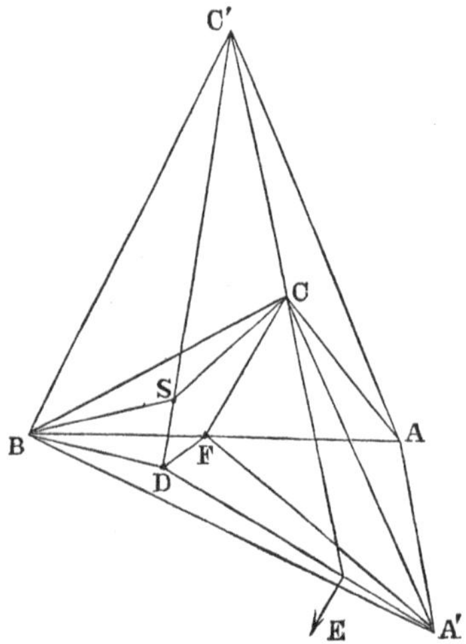

Join DF, CF; then the angle DFB is equal to DA′A [*Euc.* III., XXII.]; therefore it is equal to BCD. Hence the triangles BCA, BCF, BCD are equiangular. Hence the circle which is the locus of the vertex when the base BC is given, and the Brocard angle is equal to that of ABC, passes through the points A, F, D. Hence it coincides with the locus of A.

36. If C′, A′, B′ be three corresponding points of figures directly similar, described on the sides of the triangle ABC, and if two of the lines AA′, BB′, CC′ be parallel, the three are parallel.

37. If ABC, A′B′C′ be two co-symmedian triangles, then $\cot A + \cot A' = \cot B + \cot B' = \cot C + \cot C' = \frac{2}{3}\cot\omega$. (TUCKER.)

38. If Ω be a Brocard point of the triangle ABC, and ρ_1, ρ_2, ρ_3 the circumradii of the triangles $A\Omega B$, $B\Omega C$, $C\Omega A$; then $\rho_1\rho_2\rho_3 = R^3$. (TUCKER.)

39. If Ω be a Brocard point of a harmonic polygon of n sides, ρ_1, ρ_2, ρ_3, &c., the circumradii of the triangles $A\Omega B$, $B\Omega C$, $C\Omega D$, &c.; then $\rho_1\rho_2\rho_3 \ldots \rho_n = 2^n \cos^n \frac{\pi}{2n} R^n$.

40. If the line joining two corresponding points of directly similar figures, F_1, F_2, F_3, described on the sides of the triangle ABC, pass through the centroid, the three corresponding points are collinear, and the locus of each is a circle.

Dem.—Let D_1, D_2, collinear with the centroid G, be corresponding points of F_1, F_2, and let AN, BM, CL be the medians intersecting in G. Join ND_1, MD_2, and produce to meet parallels to D_1D_2 drawn through A, B in the points A′, B′. Now, from the construction $NA' = 3ND_1$, and $MB' = 3MD$; and since D_1, D_2 are corresponding points of F_1, F_2; A′, B′ are corresponding points; and AA′, BB′ being each parallel to D_1D_2, are parallel to one another. Again, let D_3 be the point of F_3 which corresponds to D_1, D_2; and C′ that which corresponds to A′, B′; then the lines AA′, BB′, CC′ are parallel; and since $LD_3 = \frac{1}{3}LC'$, and $LG = \frac{1}{3}LC$; D_3G is parallel to CC′. Hence D_1, D_2, D_3 are collinear; and since the loci of A′, B′, C′ are circles, the loci of D_1, D_2, D_3 are circles. These are called M‘CAY's circles. It is evident that each of them passes through two double points, and through the centroid.

41. The polar of the symmedian point of a triangle, with respect to Lemoine's first circle, is the radical axis of that circle and the circumcircle.

42. The centre of perspective of the triangles ABC, *pqr* (Section III., Prop. 2, *Cor.* 2) is the pole of the line $\Omega\Omega'$, with respect to the Brocard circle.

43. The axis of perspective of Brocard's first and second

triangle is the polar of the centroid of the first, with respect to the Brocard circle.

44. Brocard's first triangle is triply in perspective with the triangle ABC.

45. The centroid of the triangle, formed by the three centres of perspective of Ex. 44, coincides with the centroid of ABC.

46–51. In the adjoining fig., ABCD is a harmonic quadrilateral,

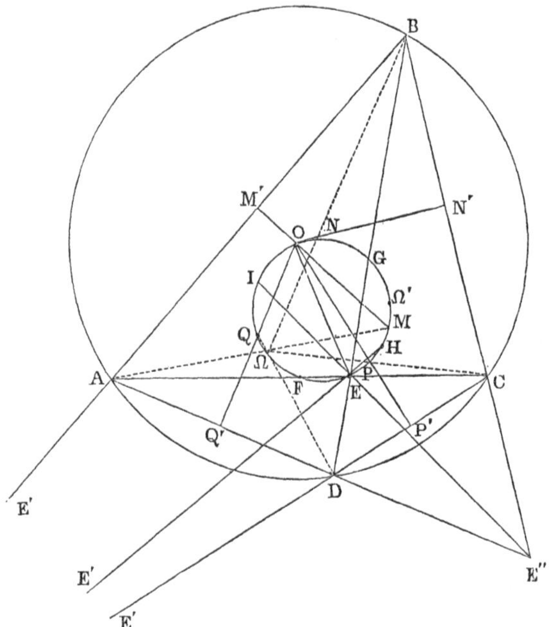

E its symmedian point, M, N, P, Q its invariable points, &c.—

1°. $\frac{PQ}{AC} = \frac{NP}{BD} = \frac{OE}{2R}$.

2°. NQ, GF, MP are parallel.

3°. The pairs of triangles MNP, FHG; MQP, GHF; NMQ, FIG; NPQ, GIF are in perspective. The four centres of per-

persctive are collinear; and the line of collinearity bisects FG at right angles.

4°. The lines AG, CG, BF, DF are tangential to a circle concentric with the circumcircle.

5°. If X, Y, Z, W be the feet of perpendiculars from E on the sides of ABCD, E is the mean centre of X, Y, Z, W.

6°. The sides of the quadrilateral XYZW are tangential to a circle concentric with the circumcircle.

52–55. If R, R′ be the symmedan points of the triangles ABC, ADC; S, S′ of the triangles BCD, DAB; then—

1°. The quadrilaterals ABCD, S′RSR′, have a common harmonic triangle.

2°. The four lines RS, S′R′, AD, BC, are concurrent.

3°. If E′ be the pole of AC; E″ of BD, the three pairs of points A, C; S′, S; E, E″, form an involution of which E, E″ are the double points.

4°. If through E a parallel to its polar be drawn, meeting the four concurrent lines AD, S′R′, RS, BC in the points λ, μ, ν, ρ, the four intercepts $\lambda\mu$, μE, Eν, $\nu\rho$ are equal; and a similar property holds for the intercepts on the parallel made by the lines AB, S′R, R′S, CD.

56. If two triangles, formed by two triads of corresponding points of three figures, F_1, F_2, F_3, directly similar, be in perspective, the locus of their centre of perspective is the circle of similitude of F_1, F_2, F_3. (Tarry.)

57. If the symmedian lines AK, BK, CK, &c., of a harmonic polygon of an odd number of sides, be produced to meet the circumcircle again in the points A′, B′, C′, &c., these points form the vertices of another harmonic polygon; and these two polygons are co-symmedian, and have the same Brocard angles, Brocard points, Lemoine circles, cosine circles, &c.

58. If three similar isosceles triangles BEA′, CAB′, ABC′ be described on the sides of a triangle ABC, prove that the axis of perspective of the triangles ABC, A′B′C′ is perpendicular to the line joining their centre of perspective to the circumcentre of ABC. (M'Cay.)

59. In the same, if perpendiculars be let fall from A, B, C on B′C′, C′A′, A′B′, prove that their point of concurrence is col-

linear with the centre of perspective and the circumcentre of ABC. (*Ibid.*)

60. If the three perpendiculars of a triangle be corresponding lines of three figures directly similar, the circle whose diameter is the line joining the centroid to the orthocentre is their circle of similitude.

61. If the base of a triangle be given in magnitude and position, and the symmedian through one of the extremities of the base in position, the locus of the vertex is a circle which touches that symmedian.

62. If through the Brocard point Ω three circles be described, each passing through two vertices of ABC, the triangle formed by their centres has the circumcentre of ABC for one of its Brocard points. (Dewulf.)

63. If through the Brocard point Ω of a harmonic polygon of any number of sides circles be described, each passing through two vertices of the polygon, their centres from the vertices of a harmonic polygon are similar to the original.

64. In the same manner, by means of the other Brocard point, we get another harmonic polygon. The two polygons are equal and in perspective, their centre of perspective being the circumcentre of the original polygon.

65. In the same case, the circumcentre of the original polygon is a Brocard point of each of the two new polygons.

66. If ρ be the circumradius of either of the new polygons, the radius of Lemoine's first circle of the original harmonic polygon is $\rho \tan \omega$.

67–72. If D_1, D_2, D_3 be corresponding points of F_1, F_2, F_3, three figures directly similar, the loci of these points are circles in the following cases :—

1°. When one of the sides of the triangle $D_1D_2D_3$ is given in magnitude.

2°. When one angle of the triangle $D_1D_2D_3$ is given in magnitude.

3°. When tangents from any two of the points D_1, D_2, D_3 to a given circle have a given ratio.

4°. When the sum of the squares of the distances of D_1, D_2, D_3 from given points, each multiplied by a given constant, is given.

5°. When the sum of the squares of the sides of the triangle $D_1D_2D_3$, each multiplied by a given constant, is given.

6°. When the Brocard angle of the triangle $D_1D_2D_3$ is given.

73. The poles of the sides of the triangle ABC, with respect to the corresponding M'Cay's circles, are the vertices of Brocard's first triangle.

74. The mean centre of three corresponding points in the system of figures, Ex. 60, for the system of multiplies a_2, b_2, c_2, is the symmedian point of the triangle ABC.

75. If from the middle points of the sides of the triangle ABC tangents be drawn to the corresponding Neuberg's circles, the points of contact lie on two right lines through the centroid of ABC.

76. The circumcentre of a triangle, its symmedian point, and the orthocentre of its pedal triangle, are collinear. (TUCKER.)

77. The orthocentre of a triangle, its symmedian point, and the orthocentre of its pedal triangle, are collinear. (E. VAN AUBEL.)

78. The perpendicular from the angular points of the triangle ABC on the sides of Brocard's first triangle are concurrent, and their point of concurrence (called TARRY'S POINT) is on the circumcircle of ABC.

79. The Simson's line of Tarry's point is perpendicular to OK.

80. The parallels drawn through A, B, C to the sides (B'C' C'A', A'B'), or to (C'A', A'B', B'C'), or (A'B', B'C', C'A') of the first triangle of Brocard, concur in three points, R, R', R''. (NEUBERG.)

81. The triangles RR'R'', ABC have the same centroid. (*Ibid.*)

82. R is the point on the circumcircle whose Simson's line is parallel to OK.

83. If from Tarry's point, Ex. 78, perpendiculars be drawn to the sides BC, CA, AB of the triangle, meeting those sides in (α α_1, α_2), (β, β_1, β_2), (γ, γ_1, γ_2), the points α, β_1, γ, are collinear (Simson's line). So also α_1, β_2, γ, and α_2, β, γ_1 are collinear systems. (NEUBERG.)

84. M'Cay's circles are the inverses of the sides of Brocard's first triangle, with respect to the circle whose centre is the centroid of ABC, and which cuts its Brocard circle orthogonally.

85. The tangents from ABC to the Brocard circle are proportional to a^{-1}, b^{-1}, c^{-1}.

86. If the alternate sides of Lemoine's hexagon be produced to meet, forming a second triangle, its inscribed circle is equal to the nine-points circle of the original triangle.

87. If K be the symmedian point of the triangle ABC, and the angles ABK, BCK, CAK be denoted by θ_1, θ_2, θ_3, respectively; and the angles BAK, CBK, ACK by ϕ_1, ϕ_2, ϕ_3, respectively; then $\cot\theta_1 + \cot\theta_2 + \cot\theta_3 = \cot\phi_1 + \cot\phi_2 + \cot\phi_3 = 3\cot\omega$. (TUCKER.)

88. If A_1, B_1, C_1 be the vertices of Brocard's first triangle, the lines BA_1, AB_1 are divided proportionally by $\Omega\Omega'$.

89. The middle point of AB, A_1B_1, $\Omega\Omega'$ are collinear (STOLL.)

90. The triangle formed by the middle points of A_1, B_1, C_1, is in perspective with ABC. (*Ibid.*)

91. If the Brocard circle of ABC intersect BC in the points M, M′, the lines AM, AM′ are isogonal conjugates with respect to the angle BAC.

92. If Ω, Ω' be the Brocard points of a harmonic polygon of n sides, $\Omega\Omega' = 2R\sin\omega\sqrt{(\cos^2\omega - \sin^2\omega \,.\, \tan^2\frac{\pi}{n})}$.

93. If the polars of the points B, C, with respect to the Brocard circle of the triangle ABC, intersect the side BC in the points L, L′, respectively; the lines AL, AL′ are isogonal conjugates with respect to the angle BAC.

94. The reciprocal of any triangle with respect to a circle, whose centre is either of the Brocard points, is a similar triangle, having the centre of reciprocation for one of its Brocard points.

95. If the angles which the sides AB, BC, CD . . . KL of a harmonic polygon subtend at any point of its circumcircle be denoted by α, β, γ, . . . λ, the perpendiculars from the Brocard points on the sides are proportional respectively to the quantities, $\sin\lambda$ cosec A, $\sin\alpha$ cosec B, $\sin\beta$ cosec C, . . . sin K cosec L, and their reciprocals.

96. The triangle ABC, its reciprocal with respect to the Brocard circle, and the triangle pqr [Section III., Prop. 2, *Cor.* 2], are, two by two, in perspective, and have a common axis of perspective.

97. If the sides AB, BC, CD, DE, &c., of a harmonic polygon of any number of sides, be divided proportionally in the points L′, M′, N′, P′, &c., the circumcircles of the n triangles L′BM′, M′CN′, N′DP′, P′EQ′, &c., have one point common to all, and each of them bisects one of the symmedian chords of the polygon.

98. The locus of the common point in Ex. 97, as the points L′, M′, N′, &c., vary, is the Brocard circle of the polygon.

99. If Ω, Ω′ be the Brocard points of a harmonic polygon of any number of sides (n); then the products AΩ . BΩ . CΩ . . . .

$$= A\Omega' \,.\, B\Omega' \,.\, C\Omega' \; \ldots\ldots = \left(R \sec \frac{\pi}{n} \sin \omega\right)^n.$$

100. If ABCDEF be a harmonic hexagon; L, M, N, P, Q, R points which divide proportionally the sides AB, BC, CD, &c.; the circles through the pairs of points L, R; M, Q; N. P, and through any common point on the Brocard circle of the hexagon, are coaxal.

101. If the lines AΩ, BΩ, CΩ meet the opposite sides in A′, B′, C′, prove that

$$\Delta\, ABC \,/\, \Delta\, A'B'C' = (a^2 + b^2)\,(b^2 + c^2)\,(c^2 + a^2) \,/\, (2\, a^2b^2c^2).$$

102. A harmonic polygon of any number of sides can be projected into a regular polygon of the same number of sides, and the projection of the symmedian point of the former will be the circumcentre of the latter.

103. The sum of the squares of the perpendiculars from the symmedian point of a harmonic polygon on the sides of the polygon is a minimum.

104. Similar isosceles triangles, BA′C, CB′A, AC′B, are described on the sides of a triangle ABC; then if ABC, and the triangles whose sides are AB′, BC′, CA′ and A′B, B′C, C′A, respectively, be denoted by F_1, F_2, F_3, the triangle of similitude of F_1, F_2, F_3 is ΩOΩ′, formed by the Brocard points and circumcentre of ABC; and their symmedian points are also their invariable points. (Neuberg.)

105. If the coaxal system, consisting of the circumcircle of a harmonic polygon of any number of sides, its Brocard circle, and their radical axis, be inverted into a concentric system, the radii of the three inverse circles are in GP.

106. If two pairs of opposite summits of a complete quadrilateral be isogonal conjugates with respect to a triangle, the remaining summits are isogonal conjugates.

107. Prove a corresponding property for isotomic conjugates.

108. In fig., p. 207, prove that

$$OS : OS' :: \cos\left(\omega - \frac{\pi}{n}\right) : \cos\left(\omega + \frac{\pi}{n}\right).$$

109. If AA′, BB′, CC′ be fixed chords of a circle X, and circles cutting X at equal angles be described through the points A, A′; B, B′; C, C′, respectively, the locus of their radical centre is a right line. (M'Cay.)

110. Find the locus of a point in the plane of a triangle, which is such that the triangle formed by joining the feet of its perpendiculars may have a given Brocard angle.

111. If the extremities of the base of a triangle be given in position, and also the symmedian passing through one of the extremities, the locus of the vertex is a circle.

112. If through the extremities A, B; B, C; C, D, &c., of the sides of a harmonic polygon circles be described touching the Brocard circle, the contacts being all of the same species, these circles cut the circumcircle at equal angles, and are all tangential to a circle coaxal with the Brocard circle and circumcircle.

113. The radical axis of the circumcircle and cosine circles of a harmonic quadrilateral passes through the symmedian point of the quadrilateral.

114. If the Brocard angles of two harmonic polygons, A, B, be complementary, and if the cosine circle of A be the circumcircle of B, the cosine circle of B is equal to the circumcircle of A.

115–117. In the same case, if the cosine circle of B coincide with the circumcircle of A; and n, n' be the numbers of sides of the polygons; ω, ω' their Brocard angles; δ the diameter of their common Brocard circle; then—

1°. $\tan\omega = \cos\frac{\pi}{n} \div \cos\frac{\pi}{n'}$.

2°. The Brocard points of A coincide with those of B.

3°. $\delta^2 \cos^2\frac{\pi}{n} = R^2 \cos\left(\frac{\pi}{n} + \frac{\pi}{n'}\right) \cos\left(\frac{\pi}{n} - \frac{\pi}{n'}\right)$.

118. If through the vertices of a harmonic polygon of any number of sides circles be described, cutting its circumcircle and Brocard circle orthogonally, their points of intersection with the

Brocard circle form the vertices of a harmonic polygon, and the Brocard circle of the latter polygon is coaxal with the Brocard circle and circumcircle of the former.

119. If the symmedian lines AK, BK, CK of a triangle meet the sides BC, CA, AB in the points K_a, K_b, K_c, respectively, the triangle ABC, its reciprocal with respect to its circumcircle, and the triangle $K_aK_bK_c$, have a common axis of perspective.

120. If DE be the diameter of the circumcircle of ABC, which bisects BC, and K_a be the intersection of the symmedian AK with BC, the line joining the middle point of BC to K meets the lines EK_a, DK_a, respectively, on the bisectors, internal and external, of the angle ABC.

121. If a heptagon circumscribed to a circle has for points of contact the vertices of a harmonic heptagon, the seven hexagons obtained by omitting in succession a side of the original poylgon have a common Brianchon point.

122. Prove $\Sigma \sin A \cos (A + \omega) = 0$.

123. If O be the circumcentre of a harmonic polygon of n sides, L one of the limiting points of the circumcircle and Brocard circle; then if $OL = R \cot \theta$,

$$\cos 2\theta = \tan \omega \tan \frac{\pi}{n}.$$

124. If ABCD, &c., be a harmonic polygon, and if a circle described through the pairs of points A, B; B, C; C, D, &c., touch the radical axis of the circumcircle and Brocard circle, the points of contact are in involution.

125. Any four given sides of a harmonic polygon meet any of the remaining sides in four points, whose anharmonic ratio is constant.

126. The quadrilateral formed by any four sides of a harmonic polygon is such, that the angles subtended at either of the Brocard points, by opposite sides, are supplemental.

127. The reciprocals of two co-symmedian triangles, with respect to their common symmedian point, are equiangular.

128. The reciprocals of two co-symmedian triangles, with respect to either of their common Brocard points, are two co-symmedian triangles.

129. The reciprocal of a harmonic polygon, with respect to either of its Brocard points, is a harmonic polygon, having the centre of reciprocation for one of its Brocard points.

130. If two circles, W, W′, coaxal with the circumcircle and Brocard circle of a harmonic polygon, be inverse to each other, with respect to the circumcircle; then the inverses of the circumcircle and the circle W, with respect to any point in the circumference of W′, are respectively the circumcircle and Brocard circle of another harmonic polygon whose vertices are the inverses of the vertices of the former polygon.

131. If R′ be the radius of the cosine circle of a harmonic polygon of n sides; Δ, δ, the diameters of its Lemoine circle and Brocard circle, respectively; then

$$\Delta^2 - \delta^2 = R'^2 \sec^2 \frac{\pi}{n}.$$

132. If the vertices of a harmonic polygon of n sides be inverted from any arbitrary point into the vertices of another harmonic polygon, the inverses of the centres of inversion of the former will be the centres of inversion of the latter.

133. The mean centre of the vertices of a cyclic quadrilateral is a point in the circumference of the nine-point circle of the harmonic triangle of the quadrilateral. (Russell.)

134. Prove that in the plane of any triangle there exist two points whose pedal triangles with respect to the given triangle are equilaterals.

135. Prove that the loci of the centres of the circumcircles of the figures F_2, F_3, Ex. 13, page 222, are circles.

136. If A′, B′, C′ be the points where Malfatti's circles touch each other, prove that the triangles ABC, A′B′C′ are in perspective.

137. Prove the following construction for Steiner's point R (Ex. 80). With the vertices A, B, C of the triangle as centres, and with radii equal to the opposite sides, respectively, describe circles. These, it is easy to see, will intersect, two by two, on the circumcircle in points A_1, B_1, C_1. Then the line joining the intersection of BC and B_1C_1 to A, will meet the semicircle in the point required.

138–140. If the perpendiculars of the triangle ABC, produced if necessary, meet the circles of Ex. 137 in the points A′, B′, C′, prove—

1°. Area of Δ A′B′C′ = 4 cot ω . Δ ABC.

2°. Sum of squares of A′B′, B′C′, C′A′

$$= 8\,\Delta\,ABC\,(2 \cot \omega - 3).$$

3°. If ω' be the Brocard angle of A′B′C′,

$$\cot \omega' = (2 \cot \omega - 3) \,/\, (2 - \cot \omega).$$ (Neuberg.)

141–144. In the same case, if the perpendiculars produced through the vertices meet the circles again in A″, B″, C″, prove that—

1°. $\Delta A'B'C' + \Delta A''B''C'' = 8\ \Delta ABC$.

2°. Sum of squares of sides of A′B′C′, A″B″C″

$$= 32\,ABC \cot\omega.$$

3°. Sum of the cotangents of their Brocard angles

$$= 2\cot\omega \,/\, (4 - \cot^2\omega). \quad (\textit{Ibid.})$$

145. Prove that the circle in Ex. 60 is coaxal with the nine-point circle and the Brocard circle.

LONGCHAMPS' CIRCLE, 146–158.—The circle which cuts orthogonally the three circles of Ex. 137 has been studied by M. LONGCHAMPS, in a paper in the *Journal de Mathematiques Speciales* for 1866. The properties which he proves both of a special nature, and also in connexion with recent geometry, are so interesting that we think it right to give some account of them here. The demonstrations are in all cases very simple, and form an excellent exercise for the student. We shall denote Longchamps' circle by the letter L, and the radical axis of it and the circumcircle by λ. It will be easily seen that the circle is real only in the case of obtuse-angled triangles.

146. The centre of L is the symmetrique of the orthocentre of ABC, with respect to its circumcentre.

147. The radius of L is equal to the diameter of the polar circle of ABC.

148. The circle L is orthogonal to the circles whose centres are the middle points of the sides of ABC, and whose radii are the corresponding medians.

149. If I, I′ be two isotomic points on any side BC of the triangle, the circle whose centre is I and radius AI′ belongs to a coaxal system.

150. The line λ is the polar of the centroid of the triangle ABC with respect to L.

151. λ is the isotomic conjugate of the Lemoine line of the triangle ABC, with respect to its sides.

152. λ is parallel to the line which joins the isotomic conjugates of the Brocard points of ABC.

153. The trilinear pole of λ, with respect to the triangle ABC,

is one of the centres of perspective of ABC, and its first Brocard triangle.

154. L intersects the circumcircle in isotomic points, with respect to the triangle.

155. The circle described with the orthocentre as centre, and radius equal to the diameter of the circumcircle, is coaxal with L and the circumcircle.

156. The circle, L, the circumcircle, and the circumcircle of the triangle A″B″C″, formed by drawing through the vertices of ABC parallels to its sides, are coaxal.

157. The centroid of ABC is one of the centres of similitude of L and the polar circle.

158. The radical axis of the circumcircle and polar circle of ABC is parallel to λ, and the centroid of ABC divides the distance between them in the 2 : 1.

INDEX.

THE END.

Printed by PONSONBY AND WELDRICK, *Dublin*.

www.ingramcontent.com/pod-product-compliance
Lightning Source LLC
LaVergne TN
LVHW091637100826
845152LV00005B/81
* 9 7 8 1 4 1 8 1 6 6 0 9 0 *